Howard W. Sams

Guide to

Satellite TV
Technology

Howard W. Sams

Guide to

Satellite TV Technology

John A. Ross

PROMPT® PUBLICATIONS

©1999 by Howard W. Sams & Company

PROMPT© **Publications** is an imprint of Howard W. Sams & Company, A Bell Atlantic Company, 2647 Waterfront Parkway, E. Dr., Indianapolis, IN 46214-2041.

International Standard Book Number: 0-7906-1176-7
Library of Congress Catalog Card Number: 98-068489

Acquisitions Editor: Loretta Yates
Editor: J.B. Hall
Assistant Editors: Kim Heusel
Typesetting: J.B. Hall
Indexing: J.B. Hall
Cover Design: Christy Pierce
Graphics Conversion: J. B. Hall
Illustrations and Other Materials: Courtesy of the Author

PRINTED IN THE UNITED STATES OF AMERICA

9 8 7 6 5 4 3 2 1

Contents

CHAPTER 3
Dish Antennas 65

CHAPTER 4
Moving From
Definition to Assembly 93

CHAPTER 5
Installing the
Feed and Amplifier 125

CHAPTER 6
An Inside Look At
Satellite Receivers 155

CHAPTER 7
Satellite Receiver
Functions and Options 187

CHAPTER 8
Digital Satellite 221

CHAPTER 9
Digitizing Satellite Signals 247

CHAPTER 10
Encryption and
Decryption Technologies 267

CHAPTER 11
Expanding Your System 297

Index 325

Preface

Satellite television technology has become commonplace since its entrance into the consumer marketplace more than fifteen years ago. As systems have progressed from the early home-made C-band dishes and crude receivers to the mass-produced C, Ku, and Direct Broadcast systems of today, consumers have become accustomed to receiving a large number of diverse programming choices. However, as the market for satellite television receiving systems continues to change and explode, the need for installation, maintenance, and service has also increased.

Howard W. Sams Guide to Satellite Television Technology provides the type of information that consumers need to make wise purchases of home satellite equipment. Moreover, the book provides practical information about installing, maintaining and repairing their satellite television reception systems. For those readers who want to know how and why the systems operate, the book also uses a number of mathematical equations to merge theory and practice.

To accomplish this, the book uses everyday language, photographs, line art, schematics, and other graphics to:

- explain how satellite technology operates

- define the differences between C, Ku, and DBS signals

- illustrate the different parts of satellite television reception systems

- show methods for installing systems at either the home or on an recreational vehicle

- list common repair techniques for mechanical equipment

- provide information about repairing electrical and electronic components, and

- describe common failure symptoms and solutions

While *Howard W. Sams Guide to Satellite Television Technology* assists consumers with the installation, maintenance, and repair of their satellite systems, it also provides the information needed to make a wise purchase. In addition, the book will contain sufficient technical content so that it will appeal to technicians as a reference. To accomplish these goals, the book covers all aspects of satellite television technology through a writing style that breaks "tech-talk" down into an easily-understood reading level. As the Table of Contents indicates, the book takes a step-by-step approach to the learning process.

When viewing the outline of the book, this step-by-step approach becomes more evident. Chapter 1 begins by providing basic information about electronic systems and components as it leads to a discussion about communications signals. The chapter provides a general overview of passive and active components that establishes a foundation for understanding how satellite receiving equipment operates. In addition, the chapter introduces key concepts such as filtering and polarization that have applications to satellite technology.

Chapter 2 continues with the basic building block approach by using definitions of amplification, oscillation, and feedback to introduce modulation and heterodyning. From there, the chapter briefly discusses how satellite signals move from the transmitter to the receiver and defines the different types of satellite signals. In addition, chapter two compares and contrasts different types of satellite systems while listing available consumer options.

Chapters 3-7 approach the installation of a C-band system in steps that involve mechanical, electrical, and electronic sub-sections. Within these steps, the chapters introduce the reader to everything from dish antenna design and construction to microwave signal amplifier and feed characteristics. Chapter 3 emphasizes the theory of reflector antennas and addresses operating characteristics while Chapter 4 talks in practical terms about the process of installing the dish antenna, ground pole and cabling.

Chapters 5, 6, and 7 cover the electronic portion of a C-band satellite system. With Chapter 5, the book takes a careful look at installing the feed and amplifier and considers topics such as polarization, down conversion, and cabling. Chapter 6 begins a two-chapter look inside satellite receivers with its overview of power supplies. The chapter begins by showing basic power supply designs and moves to schematic diagrams of power supplies used in a positioner control and receiver. Chapter 7 studies the signal processing circuitry of a satellite receiver with individual schematic diagrams of the receiver tuner, microprocessor control circuitry, and video signal processing circuitry.

Chapters 8 and 9 apply the same techniques towards Ku-band and direct transmission systems. While chapter eight provides an in-depth look at Ku-band signals, its emphasis turns to digital satellite signals and the service providers for direct broadcast satellite signals. The chapter progresses from the signal at the satellite to the conversion of signals from an analog to digital format and then to the reception of digital signals at the home. Chapter nine builds off those discussions by working through the installation of a digital satellite system.

Chapter 10 goes beyond the installation of satellite equipment with a detail look at descrambling technologies. Within the chapter, the reader will find information about video and audio signal compression and the encoding of information onto an existing signal. The chapter defines terms such as quantization, sampling, statistical multiplexing, and forward error correction while discussing the fundamentals of MPEG-2 compression.

Chapter 11 works as a capstone chapter by considering how to expand or upgrade an existing satellite television system. The opening section of the chapter introduces multi-receiver system designs and blends into the following section that covers satellite master antenna television systems. In addition to multi-receiver designs, the book also illustrates methods for installing a DBS system on a recreational vehicle. The chapter concludes by showing how specialized DBS systems can provide Internet access at the home or office. As with the other portions of the text, this section provides information geared towards both the consumer and the technician.

I thank God for giving me the knowledge and skill needed for this project. In addition, I also thank my parents—John C. and Lorraine Ross—for assisting me and for continually boosting my spirits; Mr. J.B. Hall of Prompt Publications for his patience and support; and the staff of Forsyth Library in Hays, Kansas, for their help.

CHAPTER

1

Electronic Systems and Signals

The 20th century has introduced all of us to the information age. Radio and then television broadcasts began a trend where consumers had near-instant access to entertainment, commercial advertisements, and news. This trend became even more evident as radios, televisions and the antennas needed to receive the signals became readily available and plentiful. As the demand for information access grew, corporations responded by forming broadcast networks and cable companies.

The information age has become a large part of our lives because of the availability and evolution of technologies. Along with radio and television technologies, consumers also have access to different types of personal computers and the Internet. As a result, the information age includes not only audio and video information but also data.

Satellite transmission and reception technologies exist as one particular factor in the growth and availability of the various forms of information. On a large scale, the broadcast networks and cable companies would have little influence without the benefits provided by satellite technologies. The advent of satellite technologies for the consumer has allowed businesses and individuals to access an ever-growing, broad array of information and entertainment services. In addition, the use of satellite technologies as a broadcast medium has paved the way for the convergence of different communications technologies.

Chapter one of this text introduces basic concepts that work as factors for the transmission and reception of information through satellite systems. The chapter begins by defining signals, electromagnet waves, and frequencies. From there, the chapter moves towards a general discussion of how those concepts work within the transmission and reception of information through satellite systems. Each of the topics shown within chapter one provides the foundation for a greater understanding of how the technology and the systems function.

Electronic Systems

A group of electronic components interconnected to perform a function or group of related functions makes up an *electronic system*. All electronic systems have the three basic parts shown in *Figure 1.1*. Those parts are the source, circuit, and load. The *source* is a device such as a power supply in a satellite receiver that develops a voltage or a combination of voltages.

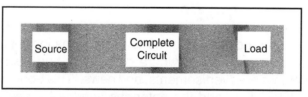

Figure 1.1. Three basic parts of an electronic system.

To take advantage of the force associated with an electric charge and use that force to produce work, we must control the flow of electrons. A load is some type of device that performs a specific function when current flows through the device. Because a load opposes current in an electric circuit, it has resistance. We can define resistance as the property of a device that opposes current in an electric circuit.

Electron flow can only occur if a complete path exists from the source to the load and back to the source. Therefore, a complete circuit must exist for electron flow. The insertion of a switch—one of the simplest of control devices—allows the control of the electron flow. A *switch* is a device that opens and closes the path for electrons.

Electronic systems divide into analog and digital circuits. An *analog* circuit has an output that varies smoothly over a given range and has an infinite number of voltage and values. Each of those values corresponds with some portion of the input. *Digital* circuits have either an "on" or an "off" output state. The application of an input signal to a digital circuit produces either of the output states but—unlike the analog circuit—no intermediate output conditions.

Source Voltages

In electricity and electronics, the term "*voltage*" describes a "difference in potential," or the amount of electric force that exists between two charged bodies. A *volt* is the standard unit of measurement for expressing the difference in potential. Because this force in volts causes the movement of electrons, it is defined as the *electromotive force*, or *emf*. Voltage is defined as either a positive or negative force with reference to another point.

If we have two bodies that have an opposite charge, or a difference in potential, and then connect those bodies with a conductor, the charged body with a more positive potential will attract free electrons from the conductor. As the free electrons move from the conductor to the charged body, the conductor assumes a positive charge because of the loss of electrons. Because of this, excess electrons in the other, more negative charged body begin to flow into the conductor. This flow of electrons is called *current* and continues as long as the difference in potential exists. The basic measure of current is the *ampere*.

The polarity of the source voltage used in a dc circuit does not change. In *direct current*, or *dc*, circuits, electrons flow in only one direction. In *alternating current*, or *ac*, circuits, the direction of electron flow changes periodically. Because of the changes in direction, the polarity of the source voltage also changes from positive to negative.

The ability of a circuit or device to pass current is defined as conductance. Depending on its size and type of material, a conductor carries electricity in varying quantities and over varying distances. A large diameter wire provides a greater surface area, has less resistance to current, and can carry more current. A material such as copper has better conductance characteristics than a material such as platinum. An inverse relationship exists between conductance and resistance. Conductance is measured in either mhos or siemens.

Passive Components

Devices with different values of resistance are connected into circuits to control the amount of current. Because resistance opposes current, the current in a circuit is always inversely proportional to the resistance in a circuit. If the resistance increases, the amount of current decreases. If the resistance decreases, the amount of current increases. *Resistance* is measured in ohms, milliohms (.001), kilohms (1000), and megohms (1,000,000). A *resistor* is a device that offers a certain amount of opposition to the flow of current. *Figure 1.2A* shows the schematic symbol for a resistor while *Figure 1.2B* provides a photograph of a resistor.

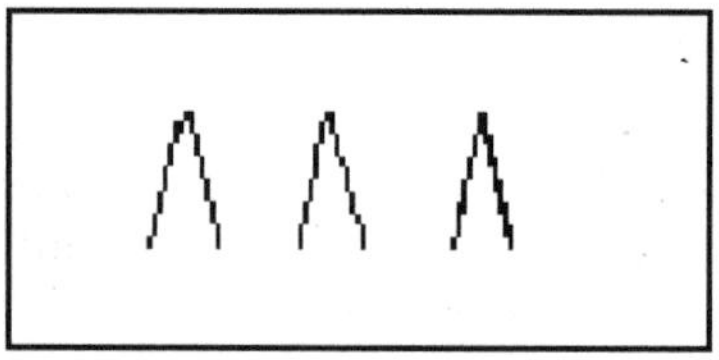

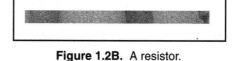

Figure 1.2B. A resistor.

Figure 1.2A. Schematic symbol for a resistor.

Capacitors store an electrical charge and act as buffers in power supplies or as filters that prevent voltage spikes from reaching sensitive solid-state electronic devices. A capacitor exists whenever two conductors are separated by a dielectric. Capacitance occurs with the separation of two or more conducting materials by an insulating material. We measure capacitance with basic units called the Farad (1), the millifarad (.001 Farad), the microfarad (.000001), and the picofarad (0.0000000000001 Farad or pF). The capacitance value depends on the amount of total surface area taken by the conducting materials, the amount of spacing between the conducting materials, and the thickness and type of insulating material. *Figure 1.3A* shows the schematic diagram for capacitors while *Figure 1.3B* shows different types of capacitors. The different types of capacitors range from mylar to electrolytic.

Figure 1.3A.
Schematic symbol for a capacitor.

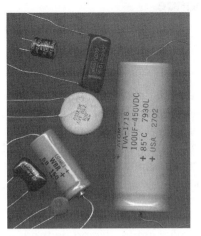

Figure 1.3B. Photograph of different capacitor types. The photograph includes electrolytic, disk and mylar capacitors.

An inductor is a conductor surrounded by a magnetic field. Inductance is a value associated with coils and is measured in a basic units called the Henry (1H), the milliHenry (.001 Henry or mH), and the microHenry (.0000001 or uH). The inductance value of a coil depends on size, number of windings, and type of core material. Small coils have lesser values than large coils while coils with a higher number of windings have larger values. When considering core materials, powered iron cores yield a higher inductance than brass or copper cores.

Inductances also have reactance and impedance values. The impedance value of an inductance varies with the value of the ac frequency and the value of the inductance. Any increase in the frequency or the inductance values also increases the inductive reactance. *Figures 1.4A* shows the schematic symbol for an inductor. *Figure 1.4B* shows a photograph of different inductor types.

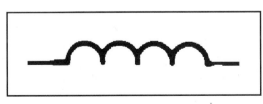

Figure 1.4A. Schematic symbol for an inductor.

Figure 1.4B. Photograph of different inductor types.

Characteristic Impedance

Impedance is a measure of resistance to current flow and varies with the value of the ac frequency and the value of the inductance. Any type of conductor—gold, copper, aluminum, or others—has a resistance to current flow that results in a loss of power. Moreover, the movement of electrical charges on the outer skin of a conductor will combine with the movement of electrical charges within the conductor to produce additional delays and losses. When we view the opposition of conductor to current flow as a combination, we can measure the characteristic impedance of the conductor as a resistance and in ohms. Manufacturers rate the cables used for video, audio, and data transmissions according to characteristic impedance.

As an example, an RG/6 coaxial cable utilized as the IF cable between the LNB and receiver of a satellite system has a characteristic impedance of 75 ohms. Older satellite systems depended on a separate low-noise amplifier and downconverter. Because of that, an additional RG-214 coaxial cable with a 50 ohm impedance was connected between the two devices.

When we assemble any type of communications system, the characteristic impedance of the transmission cables and the devices that make up the system becomes extremely important. Because every electronic component has characteristic impedance, we must ensure that the characteristic impedance of the cable matches the characteristic impedance of the component. Mismatches of characteristic impedance can occur through inadequate system designs, crimped transmission cables, or poorly constructed cables. Any mismatch in characteristic impedance will cause a signal loss to occur.

Resonance

Resonance occurs when a specific frequency causes the inductances and capacitances in either a series or parallel ac circuit to exactly oppose one another. With resonance, a single particular frequency emerges as the resonant frequency and three basic rules follow:

- Capacitors and inductors with larger values have lower resonant frequencies.

- Capacitors and inductors in series have a low impedance at the resonant frequency.

- Capacitors and inductors in a parallel circuit have a high impedance at the resonant frequency.

Traps and Filters

Electronic filters in a satellite system can eliminate noise and many types of terrestrial interference. Satellite receivers utilize different types of filters while processing the signal to ether permit the passage of desired signals or to either eliminate or bypasse unwanted signals. Low-pass filters pass all frequencies up to a selected frequency and eliminate all signals above the selected frequency. High-pass filters pass all signals above a selected frequency and eliminate all signals below the selected frequency. *Figure 1.5* uses waveforms to illustrate the differences between low-pass and high-pass filters.

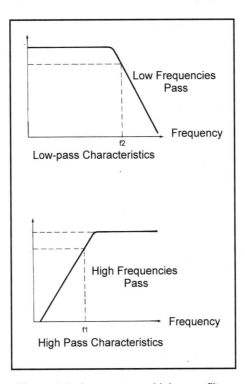

Figure 1.5. Low-pass vs. high-pass filter characteristics.

As a system processes signals, translates signals from one frequency band to another, or mixes the signals to produce usable information, filters also remove undesired signals. In some circuit designs, the removal of the undesired signals occurs through the use of sharply tuned circuits, called wave traps. These series or parallel tuned either reject, absorb, or provide a path to ground for a particular signal.

In some circuit section designs, attenuation of the undesired signals occurs through the use of *wave traps*. These series or parallel tuned circuits work within the IF interstage coupling circuits and either reject, absorb, or provide a path to ground for a particular signal. The tuned signal may be an adjacent signal, or desired signal. Depending on the requirements, wave traps fall under four categories:

> 1) Series resonant traps
> 2) Parallel resonant traps
> 3) Absorption traps, and
> 4) Bridged-T or notch traps

Attenuation of undesired signals occurs because of the reduction of signal gain at the trap frequencies. Because of the rejection of some part of the IF signal, the tuning of the traps affects the shape of the frequency response curve. Each trap affects the IF response at the edges of the passband by cutting into the skirt of the response curve.

In *Figure 1.6*, L1 and C1 make up a series resonant wave trap that shunts an unwanted adjacent video signal to ground. This shunting effect occurs because of the low impedance presented by the components at a resonant frequency of the unwanted signal. Referring to *Figure 1.7*, L2 and C2 make up a parallel resonant trap that also couples the IF signal to the 2nd IF stage. In this configuration, the components present a high impedance to any unwanted signals and prevent those signal from traveling to the amplifier.

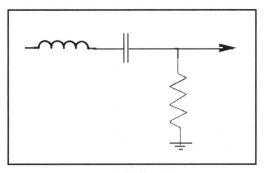

Figure 1.6. Schematic diagram of a series
resonant wave trap.

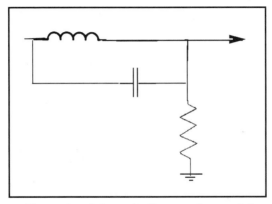

Figure 1.7. Schematic diagram of a parallel
resonant wavetrap.

As the name implies, an absorption trap absorbs energy at a resonant frequency. An absorption trap is a parallel resonant circuit that is inductively coupled to another circuit. During operation, the absorption trap absorbs energy from the circuit and attenuates the specified signal.

A bridged-T Trap blocks all frequencies that fall within its bandwidth. *Figure 1.8* shows a block diagram of a bridge-T filter and a corresponding frequency response curve. Looking at the figure, the notch filter consists of a low-pass filter, a high-pass filter, and a summing amplifier. While the high-pass filter establishes the lower cut-off frequency, or f_1, for the complete filter, the low-pass sets the upper cut-off frequency, or f_2, for the complete filter.

Moving to the response curve, the gap between the two cutoff frequencies sets the bandwidth for the filter. With an input frequency lower than f1, the input voltage produced at the output of the high pass filter equals zero while a positive voltage appears at the output of the low pass filter. With an input frequency higher than f2, the input voltage produced at the output of the low pass filter also equals zero while a positive voltage appears at the output of the high pass filter.

Thus, with either an input frequency below f1 or an input frequency higher than f2, the signal passes through to the summing amplifier and the amplifier has an output signal. However, when the input frequency fits anywhere in-between the upper and lower cut-off frequencies, neither the low-pass, the high-pass filter, or the summing amplifier has an output. As *Figure 1.8* shows, the notching action of the complete filter circuit either reduces or rejects any frequency fitting within the circuit bandwidth.

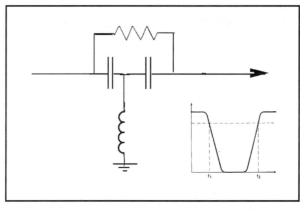

Figure 1.8. Schematic drawing of a bridged T-wave trap.

We can also use the filter circuit shown in the figure to make a transition from purely trapping circuits to circuits that provide both attenuation and coupling. Because the bandpass filter consists of two tuned circuits coupled in such a way that the circuit has a maximum response at two frequencies, it is also called a coupled filter. In the figure, both the primary and the secondary of a double-tuned coupling transformer tune to the same frequency. As we will see in the next section, the close coupling of the two coils provides the desired frequency response.

Bandpass Filters

In addition to wave traps, electronic circuits also utilize *bandpass filters* that eliminate any frequency outside of established band limits. By employing bandpass filters, the circuit assures the maximum response to a given frequency. Bandpass filter characteristics allow any frequency below the value of a specified frequency and any frequencies above the value of a specified frequency to pass. The filter blocks any frequencies above and below the specified frequencies.

Shown in *Figure 1.9*, the most simple bandpass filter consists of several tuned circuits and a terminating resistor. With this basic design, the frequency limits of the passband and a given undesired frequency that lies outside the band limits determine the amount of inductance, the amount of capacitance, and the value of the terminating resistor. Because the bandpass filter consists of two tuned circuits coupled in such a way that the circuit has a maximum response at two frequencies, it is also called a coupled filter. In the figure, both the primary and the secondary of a double-tuned coupling transformer tune to the same frequency.

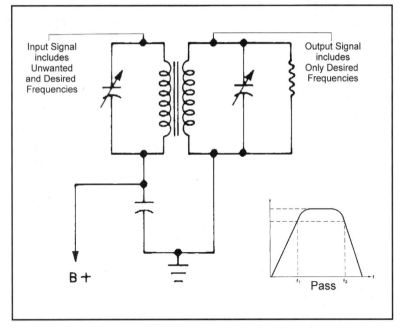

Figure 1.9. A simple bandpass circuit.

Active Components

While passive components provide resistance, inductance, and capacitance, active components such as diodes, transistors, and integrated circuits manipulate voltage and current through switching and amplification. Active components are semiconductors, or devices consisting of materials that neither fully conduct nor fully resist electricity. A diode is a two-terminal semiconductor device that conducts under specific operating conditions. Depicted in *Figure 1.10*, PN junction diodes are constructed of a more negative n-type material at the cathode and of a more positive p-type material at the anode.

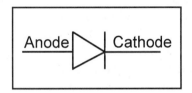

Figure 1.10. Schematic symbol
for a PN junction diode.

An ideal diode acts as an open switch when reverse biased and as a closed switch when forward biased. The term *bias* refers to the no-signal dc operating voltage or current between two of the elements of a semiconductor device. When forward biased, a more positive voltage applies to the positive p-type material than to the negative n-type material. Reverse biasing applies a more negative voltage to the positive p-type material than to the negative n-type material.

A *zener diode* is a special type of diode that operates when reverse biased. Because of this, a zener diode maintains a relatively constant voltage despite any variations in the diode current. Zener diodes work well as voltage regulators. *Figure 1.11* shows the schematic symbol for a zener diode.

Figure 1.11. Schematic symbol for a
zener diode.

Transistors are remarkable devices that can amplify small AC signals or switch a device from its on state to an off state and back. Any electronic system will have at least one transistor performing a task. To name a few of those tasks, modern electronic equipment uses transistors for signal transmission, video and audio signal reproduction and voltage regulation.

As with the diodes, transistors begin as a single crystal of silicon or germanium. The construction of the transistor features the same techniques used in diode construction—the joining of p- and n-types of semiconductor material. However, the transistor contains three sections of semiconductor material that results in two PN junctions and three terminals.

Two fundamental types of bipolar junction transistors exist. One type, called the NPN transistor, consists of a p-type material sandwiched between two n-type materials. Conversely, a PNP transistor consists of an n-type materials sandwiched between two p-type materials. We refer to transistors as bipolar junction transistors or BJTs because the devices contain both the N- and P-type materials. Referring to *Figure 1.6*, we can label the three regions of the bipolar transistor as the emitter, base and collector. The two junctions of the transistor are called the collector-base or collector junction and the emitter-base or emitter junction. The base section is common to both junctions.

Like diodes, transistors will not function without a source of energy. In most cases, this energy takes the form of an applied DC voltage. Depending on the task given the transistor, various methods of applying the DC voltages exist. In a normally biased transistor, the emitter junction is always forward biased while the collector junction is always reverse biased.

Figure 1.12 shows the cross section of the NPN transistor and its schematic symbol. This three region device has a layer of n-type material used as the emitter, a layer of p-type material as the base and a second layer of n-type material as the collector.

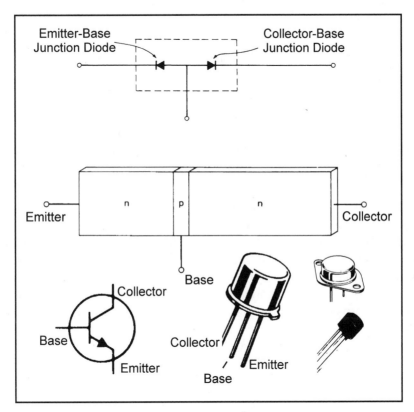

Figure 1.12. Block diagram of a transistor.

Separating the emitter from the collector, the thinner base region controls the number of charge carriers leaving the emitter. By far the largest region, the collector often works as the output section of the transistor. *Figure 1.13* shows a PNP transistor and its schematic symbol. The PNP transistor has a p-type emitter, an n-type base and a p-type collector. Considering any applied voltages, the polarities simply reverse.

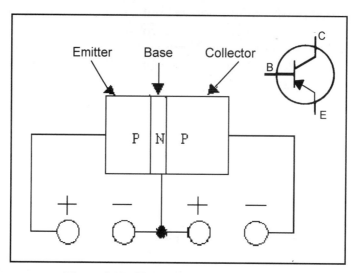

Figure 1.13. Diagrams of a PNP transistor.

Cutoff, Saturation and Breakdown

Figure 1.14 shows how the base and collector currents flow in the emitter region. Given its greater value, a heavier line represents the collector current. Even though the base current makes up only a small fraction of the entire transistor current, it plays a major role in transistor operation. With no base current, collector current cannot exist. Inside the transistor, the resistance from emitter to collector increases to a very high level. While the full supply voltage appears across the transistor, no voltage appears across the load resistance. Effectively, the transistor becomes "turned off" or enters a cutoff condition.

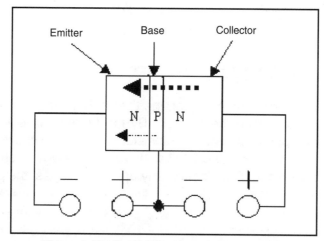

Figure 1.14. Current flow in an NPN transistor.

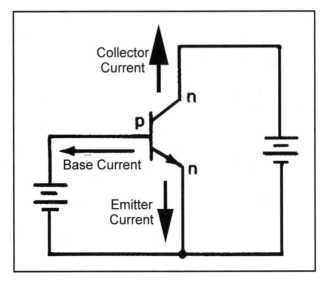

Figure 1.15. Current flow of a saturated NPN transistor.

Sometimes, another condition can draw off enough base current so that the emitter-collector resistance decreases. We call this condition *saturation*. During saturation, any further increase in base current will not cause an increase in collector current. *Figure 1.15* shows the electron flow for a saturated NPN transistor. Little or no voltage appears across the transistor.

A bipolar-junction transistor can operate as a switch by driving the device back and forth between saturation to cutoff. Switching transistors have a delay between each input transition and the time when the output transition begins. The output transition for a switching transistor takes a given amount of time to occur. Because transistor switches have internal resistance and leakage resistance, the device never completely restricts the flow of current or allows all current to pass. Capacitance found within transistors prevents the device from switching instantaneously. Usually, a transistor switch will have a common-emitter configuration because of the current and voltage gain characteristics given by that configuration.

By careful chemical composition and arrangement, it is possible to create a very small transistor directly on a layer of silicon, using various technologies to manipulate the material into the correct form. These transistors are small, fast and reliable, and use relatively little power. In a very basic sense, an *integrated circuit* such as the one shown in *Figure 1.16* is a group of transistors manufactured from a single piece of material and connected together internally, without extra wiring. Integrated circuits are also called ICs or chips and contain passive and active elements while performing a complete circuit function or a combination of complete circuit functions.

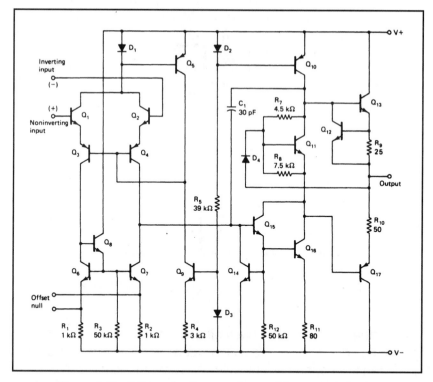

Figure 1.16. Schematic diagram of a small integrated circuit.

After the invention of the integrated circuit, it took very little time to realize the tremendous benefits of miniaturizing and integrating larger numbers of transistors into the same integrated circuit. The use of additional transistor switching allowed the implementation of more complicated functions. To accomplish this, manufacturers miniaturized components to a much greater extent and integrated large numbers of transistors while increasing hardware speed and managing power consumption and space requirements. Large-scale integration refers to the creation of integrated circuits that contain hundreds of transistors.

IC packages may house anywhere from a dozen to millions of individual components. Because computers and microprocessor control systems require hundreds of thousands of those functions, cost and size savings are critical. Integrated circuits provide the most efficient method of packaging those functions within a small area. Standard digital ICs are available for every conceivable logic application. Along with those benefits, integrated circuits also offer increased reliability and higher operating speeds.

After the invention of LSI integrated circuits, integration and miniaturization technologies continually improved and allowed the manufacture of smaller, faster and cheaper chips. Very large scale integration builds millions of transistors onto one IC and as a result packs more logic into a single device. With VLSI technology, the functions once performed by several different logic ICs became enclosed within one package.

Analog Signals

Signals can be found at transmission points, may be generated within electronic systems, and may take different forms. One of the most basic signal waveforms is the sinusoidal wave, or sine wave. Other basic types include the rectangular wave, the triangle waveform, the sawtooth waveform, and rectangular pulses.

Sine Waves

The *sinusoidal wave*, or *sine wave*, shown in *Figure 1.17A* represents a mathematical relationship of an alternating voltage or current produced by an alternator, inverter, or oscillator. When a sine wave goes above and below the zero line twice, each combination of maximum positive and negative values equals one *cycle*. In turn, each cycle subdivides into two alternations, or one-half cycle. We can view an *alternation* as the rise and fall of voltage or current in one direction.

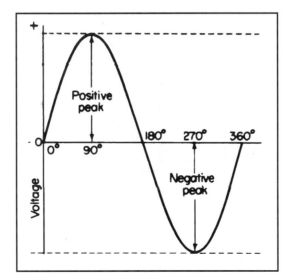

Figure 1.17A. Sine wave.

Rectangular Waves

Any waveform consisting of high and low dc voltages has a pulse width and a space width and is defined as a rectangular waveform. The *pulse width* is a measure of the time spent in the high dc voltage state while the *space width* is the measure of the time spent in the low dc voltage state. Adding the pulse width and space width together gives the cycle time or duration of the waveform. The values for the pulse width and space width are always taken at the halfway points of the waveform.

As shown in *Figure 1.17B*, a *square wave* is a rectangular waveform that has equal pulse width and space width values. While the pulse width and space width for a square wave equal one another, the opposite is true for pulses. We can define a *pulse* as a fast change from the reference level of a voltage or a current to a temporary level and then an equally fast change back to the original level. Shown in *Figure 1.17C*, pulses usually appear in a series called a train and are measured in terms of pulse repetition rate and repetition time period.

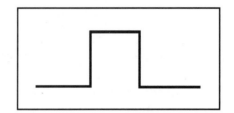

Figure 1.17B. Square wave.

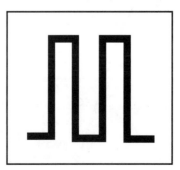

Figure 1.17C. Pulses.

Ramp, Triangular, and Sawtooth Waves

Although square waves and pulses provide the basis for modern digital communications, other types of waveforms also exist. Illustrated in *Figure 1.17D*, a *ramp waveform* has a slow linear rise and a rapid linear fall. In contrast, the *triangular waveform* shown in *Figure 1.17E* rises and falls at a constant rate and has a symmetrical shape. The sawtooth waveform shown in *Figure 1.17F* appears similar to the triangular wave but has a longer rise time and a shorter fall time.

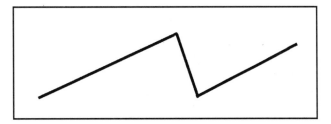

Figure 1.17D. Ramp waveform.

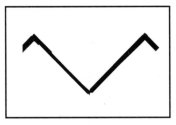

Figure 1.17E. Triangular waveform.

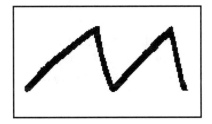

Figure 1.17F. Sawtooth waveform.

Digital Signals

In brief, an analog system operates with continuous waveforms over a given range such as the sine wave shown in *Figure 1.17A*. By definition, analog information is a quantity that may vary over a continuous range of values. Digital information is data that has only certain, discrete values. In contrast to analog information, the digital information shown in *Figure 1.18* offers precise values at specific times and changes step by step.

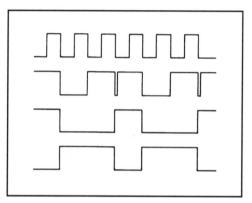

Figure 1.18. Digital information.

When considering electronic systems, this preciseness translates into immunity against electrical noise distortion and variations in component values. Digital information consists of data held in one of two states: low and high. Noise and component variations do not cause the low and high states of a digital signal to appear as opposite values.

Binary and BCD Number Systems

Computer systems work with the *binary number system*, a system that allows only two values: zero and one. The use of binary numbers in those systems breaks information down into elementary levels. In addition, because the binary number system only relies on zeroes and ones, it provides a very basic method for counting and accumulating values.

If we look at the digits of a binary number, each value in the columns equals a value based on the powers of two. As an example, a binary number represented by 111 has $2^0 = 1$ or 1x1 for the rightmost column, $2^1 = 4$ or 2x1 for the middle column, and $2^2 =$

8 or 4 x 1 for the left most column. In the decimal system, this value would equal 1x1+1x2+1x4 or 7. A binary number of 1101 equals 1x1+0x2+1x4+1x8 or a decimal equivalent of 13.

As mentioned earlier, an electronic system uses high and low signals to represent binary numbers. Each high and low signal is separated by an area of voltage that has no binary meaning. While a high signal has a value of 3 to 5VDC, a low signal has a value of 0 to 1VDC.

Bytes and Bits

The simplicity of the binary system allows computer systems to move numbers from one part of a system to another and to work with large numbers. Each binary position is called a *bit* while a group of eight bits is called a *byte*. The sum of bits provides a method for assigning a value. As a result, the number of bits required to complete a task depends on the magnitude of the number. With each bit existing as either a 1 or 0, the value of each successive bit can increase by a maximum value of 2. As a result, the individual bits of a binary number translate to the following decimal values:

32	16	8	4	2	1	1/2	1/4	1/8	1/16
2^5	2^4	2^3	2^2	2^1	2^0	2^{-1}	2^{-2}	2^{-3}	2^{-4}

Going back to the addition of binary numbers, the sum of true values in a byte equals a decimal value. That is, a byte that appears as 11001101 has an equivalent decimal value of 206.

Binary-Coded Decimal

Another counting system called *binary-coded decimal,* or *BCD*, combines the efficiency of the binary system with the familiarity of the decimal system. Keyboards, LED displays, and switches rely on a variation of the binary system called binary-coded decimal, or BCD. The BCD system uses 4 bits to represent each digit of a decimal number. As an example, decimal number 759 uses 0111 for the 7, 0101 for the 5, and 1001 for the 9 and appears as 0111 0101 1001.

Digital Building Blocks

Four basic logic elements establish most of the functions seen in electronic systems. Those logic elements are amplifiers, gates, switching elements, and delay elements. Each element follows a specific set of rules and has a specific schematic symbol as shown in *Figure 1.19*. Amplifiers increase the value of a signal as it travels from input to output. When working with digital circuits, amplifiers drive digital pulses to the next location.

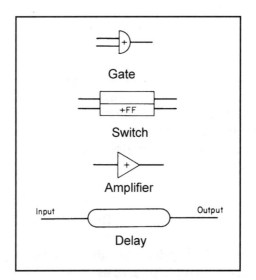

Figure 1.19. Logic symbols.

Integrated circuits called *gates* perform fundamental Boolean logic functions and combinations of the logic functions. A gate "sees" the logic values at its inputs and produces the output that corresponds with the correct logic function. Gates can be connected together so that the output of one becomes the input for another. The connecting of the gates provides more complicated Boolean logic functions and is called *combinational logic*.

An IC with less than 12 gates fits within a category called small-scale integration while ICs with between 12 and 100 gates are called medium-scale integration. Modern electronic systems utilize both the simplicity of small- and medium-scale integration and the flexibility offered by the millions of gates seen with very large-scale integration or VLSI.

With a large number of gates connected for combinational logic tasks, propagation delays can occur and, if significantly large, can cause false results in a system. This occurs because signals passed by the gates arrive at slightly different times in different parts of the system. Every digital system includes functions called enable and inhibit to ensure that the propagation delays do not accumulate and the signals remain synchronized.

The inputs for gates occur at instants in time. In addition, a gate IC does not form any type of memory. To move past those obstacles, we can connect gates into different configurations so that an input pulse will cause the gates to move from one logic state to the other. The next input signal causes the states to reverse. This configuration forms a *bistable multivibrator*, or *flip-flop*. The output of one gate in a flip-flop is the input for the other gate.

A *switching element* causes the logic state found at the output of a device to switch, or change to the opposite condition. With this, an output that originally has a logic 1 will change to a logic 0. In digital electronics, switching elements take the form of flip-flops, astable and monostable multivibrators, and Schmitt triggers. A *delay element* establishes a time delay between the input and output signals.

Analog-to-Digital and Digital-to-Analog Conversion

Many system applications require the conversion of analog signals to a digital format or the conversion of digital signals to an analog format. ADC and DAC circuits perform those operations while acknowledging requests from microprocessor controllers and other circuits. In addition to the analog/digital conversion, those circuits also require the conversion between digital formats and the translation of digital data to a display by encoders and decoders. *Analog-to-digital-conversion*, or *ADC*, circuits take an analog signal and convert the signal to a digital number that corresponds to the value of the analog signal. *Digital-to-analog conversion*, or *DAC*, circuits produce a dc output voltage that corresponds to a binary code and convert digital properties to analog voltages.

Communications Signals

Generally, electronic equipment used for communications operates with electrical energy that takes the form of electromagnetic waves. That energy may take the form of radio waves, infrared light, visible light, ultraviolet light, x-rays, and other forms. Magnetic and electric fields placed at right angles to each other and at right angles to their direction of travel make up *electromagnetic waves*. The wave-like nature of those fields becomes apparent as the magnetic and electric fields vary continually in intensity and periodically in direction at any given point.

Each complete series of variations forms a wave. As one wave travels through space, another wave immediately follows. The term *frequency* describes the number of waves that pass a point each second and the rate of polarity change. Frequency is measured in hertz. The distance in space from any given point or condition in one wave to the corresponding point of the next wave is defined as *wavelength*. Mathematically, wavelength, or l appears as:

$$l = c / f$$

where:

c = the speed of light (2.998 x 108 meters per second), and
f = frequency

Because radio waves travel through space at the speed of light, or 300,000,000 meters per second, the velocity always equals either 300,000,000 meters per second or 984,000,000 feet per second. As a result, we can measure wavelength either in meters or feet per second. During time, we have gained the ability to produce and use radio waves for the purpose of communicating over long distances.

Frequency Spectrums and Polarization

Electromagnetic wave frequencies cover a wide range that extends from the longest radio waves to the very short waves called cosmic waves. In terms of frequency, this range extends from 10 kHz at the low end to 10^{20} kHz at the high end. Of this range, or *spectrum*, radio frequencies cover a range from 10 kHz to 3×10^8 kHz.

Along with having frequency ranges, electromagnetic waves also have polarization. As with electrical wiring, electromagnetic wave polarization exists through the direction of electrical and magnetic fields. Two types of polarization occur. When the electrical and magnetic fields of a received signal remain within the same plane of the electrical and magnetic fields of the transmitted, linear polarization occurs. Horizontally polarized waves travel along the horizontal plane while vertically polarized waves travel along the vertical plane. *Figure 1.20* uses a diagram to depict linear polarization.

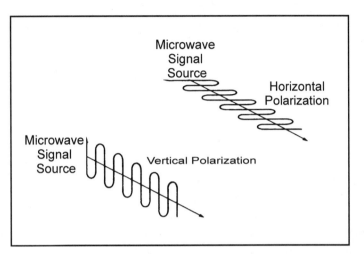

Figure 1.20. Representation of linear polarization.

Circular polarized electromagnetic waves have electrical and magnetic fields that rotate in a circular motion. In this case, a set of electrical and magnetic fields that rotate in a right-hand spiral from the satellite have right-hand circular polarization, or RHCP. Electrical and magnetic fields rotating in a left-hand spiral from the satellite have a left-hand circular polarization. *Figure 1.21* illustrates how circular polarization would appear.

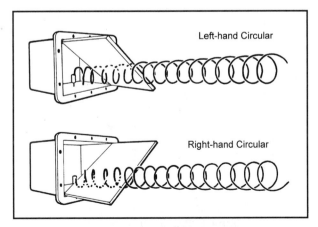

Figure 1.21. Representation of circular polarization.

Signals

In terms of electronics, we can define a *signal* as a voltage or current that has deliberately induced, time-varying characteristics. A signal voltage or current is different than a source voltage or current for several reasons. Every electrical signal has a distinctive shape described in terms of the *time domain*, or the amplitude of the signal as a function of time, and *frequency domain*, or the magnitude and relative phase of the energy.

Phase

In addition, we describe signals that have the same frequency and shape in terms of phase, or when the repetitions of the signals occur in time. In-phase signals have repetitions occurring at the same time while out-of-phase signals are displaced along a time axis. In an amplifier, the input and output have phase relationships that vary with the configuration of the amplifier circuit. In some amplifier circuits, the output voltage becomes more negative as the input voltage becomes more negative and then becomes more positive when the input voltage becomes more positive. With the output and input voltages in step with one another, no phase shift occurs. *Figure 1.22A* shows an example of in-phase signals.

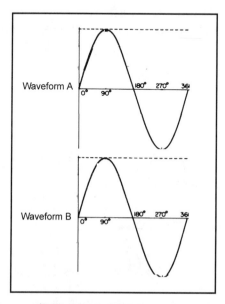

Figure 1.22A. In-phase signals.

In other amplifier circuits, the output voltage becomes more negative as the input voltage becomes less negative and less negative as the input voltage becomes more negative. With this, the output voltage signals are out-of-phase with the input voltage signals. *Figure 1.22B* shows an example of two out-of-phase signals.

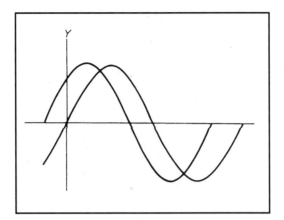

Figure 1.22B. Out-of-phase signals.

Signal Bands

Radio frequency, or *RF*, signals occupy the frequency range between 10^4 hertz and 10^{11} hertz in the frequency spectrum. As with the low frequency and dc energy seen with supply voltages, we can transmit radio frequency energy over wires. However, unlike the low frequency energy, we can also transmit radio frequency energy through space in the form of electromagnetic waves.

Certainly, many modern point-to-point transmissions of radio frequencies involve the use of satellite television transmission and reception antennas that send and receive signals through the upper atmosphere and space. Yet, we may also consider that conventional RF transmissions break down into three fundamental categories. While we can define any portion of a radio wave that travels along the surface of the earth as a *ground wave*, we call the portion of the wave radiated at an angle greater than horizontal the *sky wave*. Radio waves that travel from one antenna to another without the affect of the ground or the upper atmosphere are called *direct waves*.

The Radio Frequency Spectrum

Shown in *Table 1.1*, the RF spectrum divides into the eight categories. When considering the uses of each band, it is important how the frequencies differ in wavelength. As an example, we could transmit frequencies within the EHF band across very long distances because of the extremely long wavelength seen with those frequencies. In practice, the earth would absorb much of the power from the frequency waves. Even with those power losses, EHF transmitters still have a great range.

Designation	Abbreviation	Frequency Range (kHz)
Extremely High Frequency	EHF	30,000,000 to 300,000,000
Super High Frequency	SHF	3,000,000 to 30,000,000
Very High Frequency	UHF	300,000 to 3,000,000
High Frequency	HF	3000 to 300,000
Medium Frequency	MF	300 to 30,000
Low Frequency	LF	30 to 300
Very Low Frequency	VLF	10 to 30

Table 1.1. Radio frequency bands.

Frequencies in the LF band are even more prone to losses because of earth absorption. Yet the shorter wavelengths seen with those frequencies allow the use of highly efficient antennas. When considering the MF band, note that the commercial AM broadcast band—533 kHz to 1605 kHz—lies within those frequencies. Because of the wavelength of those frequencies, most AM radio transmissions take the form of ground waves.

The High Frequency Band covers frequencies used by foreign broadcast stations and amateur radio stations. Given the frequency band, ground wave transmissions have a limited range of 10 to 20 miles. However, sky wave transmissions between 30 and 60 MHz have a much greater range.

Frequencies found in the *VHF band* cover the VHF television bands, or 54 MHz to 72 MHz, 76 MHz to 88 MHz, and 174 MHz to 216 MHz. In addition, the VHF band includes frequencies used for commercial FM broadcast transmissions. Most conventional VHF transmissions involve direct wave transmissions. The *UHF band* covers UHF television band frequencies, or 470 MHz to 890 MHz. Conventional UHF transmissions also involve direct wave transmissions. The last band of frequencies—the SHF band—have extremely short wavelengths and are recognized as *microwave frequencies*.

Noise

When we consider the use of signals to communicate, we can define any type of disturbance other than the desired signal as *interference*, or *noise*. These extraneous signals may take a variety of forms can affect both the transmission and reception of broadcast signals. The ability of a system to reject noise is defined in terms of *signal-to-noise ratio*. A system with a high signal-to-noise ratio has a greater ability to reject noise.

Types of Noise

Atmospheric Noise Radio-wave disturbances, such as lightning, that originate in the atmosphere.

Common-mode Interference Noise caused by the voltage drops across wiring.

Conducted Interference Interference caused by the direct coupling of noise through wiring or components.

Cosmic Noise Radio waves generated by extraterrestrial sources.

Crosstalk Electrical disturbances in one circuit caused by the coupling of that circuit with another circuit.

Electromagnetic Interference EMI refers to noise ranging between the sub-audio and microwave frequencies.

Electrostatic Induction Noise signals coupled to a circuit through stray capacitance.

Hum Electrical disturbance at the power supply frequency or harmonics of the power supply frequency.

Impulse Noise Noise generated by a dc motor or generator. Impulse noise takes the form of a discrete, constant energy burst.

Magnetic Induction   Noise caused by magnetic fields.

Radiated Interference   Noise transmitted from one device to another with no connection between the two devices.

Radio-frequency Interference   RFI occurs in the frequency band reserved for communications.

Random Noise   An irregular noise signal that has instantaneous amplitude occurring randomly in time.

Static   Radio interference detected as noise in the audio frequency, or AF, stage of a receiver.

Terrestrial Interference   Unwanted earth-based communication signals.

Thermal Noise   Random noise generated through the thermal agitation of electrons in a resistor or semiconductor device.

White Noise   An electrical noise signal that has continuous and uniform power.

RF Signals and Component Properties

Circuits used to amplify and process radio frequencies differ from other circuits used for lower frequency voltages and currents because of the properties associated with the components that make up the circuits.   As we have seen, those properties are resistance, inductance, and capacitance.

For example, any RF current flowing through a conductor creates a changing magnetic field around the conductor.   When the field builds and collapses, flux lines cut the conductor.   From this, a counter EMF is induced that has a directly proportional relationship to frequency.   The *counter EMF*, or inductive reactance, opposes the current in the conductor.   Therefore, components may have a low inductive reactance at low frequencies and a high inductive reactance at the higher RF frequencies.

Another characteristic of RF signals called skin effect originates because of an alternating current flowing through a conductor is defined as the increasing resistance of a conductor at very high frequencies.   The flow of the ac current through the conductor causes magnetic flux lines to move from the center of the conductor and then expand outward. With the lines concentrated more at the center of the conductor than near the its surface, the counter EMF at the center is also greater. As a result, current concentrates along the surface of the conductor. In turn, the effective cross-sectional area of the conductor decreases. The skin effect results because the resistance of the conductor is inversely proportional to its cross-sectional area.

Radio frequencies can cause resistors to exhibit the properties of resistance, inductance, and capacitance. Inductance occurs because of the connecting leads and the resistance of the device. Capacitance occurs because the RF signal causes the resistor to have two conductive points separated by an insulator.

Like resistors, capacitors also exhibit the properties of resistance, inductance, and capacitance. Resistance exists because of the resistance of the capacitor plates, the connecting leads, and the dielectric. Inductance occurs because of the connecting leads and the dielectric plates. The inductance and capacitance form a series resonant circuit.

When the applied signal frequency has a value less than the resonant frequency of the capacitor, the capacitive reactance is higher than the inductive reactance. Thus, at low frequencies, a capacitor presents resistance and capacitance in series. When the applied signal frequency has a value higher than the resonant frequency of the capacitor, the inductive reactance is greater than the capacitive reactance. Therefore, at high frequencies, a capacitor presents inductance and resistance in series.

Inductors also offer resistance, capacitance, and inductance. Resistance exists because of the resistance found within the conductor used to wind the inductor. Because inductors have a greater ac resistance than dc resistance, inductors are prone to skin effect. Distributed capacitance in an inductor results from the many small capacitances existing between the turns of an inductor.

The inductance and distributed capacitance of an inductor form a parallel resonant circuit. At frequencies lower than the resonant frequency, an inductor has a higher capacitive reactance and a lower inductive reactance. As a result, the inductor acts like an inductance in series with a resistance. At frequencies higher than the resonant frequency, an inductor has a higher inductive reactance and a lower capacitance reactance. Therefore, the inductor acts like a capacitance in parallel with a resistance.

Resistance always dissipates energy in the form of heat. Reactance returns absorbed energy to the circuit without loss. In an inductor, the loss of energy is proportional to the effective resistance. The ratio of reactance to effective resistance is defined as the quality, or Q, of the inductor and measures the efficiency of the inductor. The mathematical representation of Q is:

$$Q = X_L / R$$

where:

Q equals quality,
X_L equals inductive reactance in ohms, and
R equals the effective resistance in ohms

This method of measuring inductor efficiency is important because Q remains constant over a wide range of frequencies. Both reactance and effective resistance increase with frequency. An inductor or circuit that has a higher value of Q has greater efficiency.

Safety First!

Whenever you work with an electronic system, a potentially deadly shock hazard exists. Your body establishes a conducting path for electricity that leads through your heart. Even if the electrical shock does not kill, it can cause you to jerk involuntarily and come into contact with sharp edges, fragile picture tubes, or other electrically live parts. As a result, the reaction to the shock could cause damage to the equipment undergoing service.

When working with electronic systems, follow these simple guidelines:

- Wear rubber-bottom shoes

- Never wear metallic jewelry that could contact circuitry and conduct electricity

- Never rest both hands on a conductive surface

- Perform as many tests as possible with the power off and the equipment unplugged

- Never work on an electronic system when tired and

- Connect and disconnect test equipment leads with all equipment powered down and unplugged. Use clip leads when accessing difficult-to-reach locations.

Summary

Chapter one defined an electronic system, highlighted the need for source voltages throughout the system, and described the ingredients that make up a typical power supply. The chapter then considered electromagnetic waves, signals and relationships between them. It then moved on to waveforms, frequency bands, harmonics, and different types of signal noise.

During this overview, the chapter also defined passive components such as resistors, capacitors, and inductors along with active components such as diodes, transistors, and integrated circuits.

Next, the emphasis shifted to radio frequency signals and the affect of those signals on different types of components. You found that RF signals affect components such as capacitors, inductors, and resistors in varying ways including skin effect and resonance. Chapter one also provided an overview of terms and concepts used in digital electronics and moved through binary numbering systems. From there, the chapter illustrated the basic building blocks of digital circuits through discussions about gates, amplifiers, switching elements, and delay elements.

CHAPTER

(2)

Moving from Signals to Satellite Systems

Chapter two builds on the fundamental information presented in chapter one by describing how electronic systems manipulate signals for communication. Within this description, the chapter defines key concepts such as amplification, oscillation, modulation, heterodyning, and demodulation. As your knowledge about satellite systems increases, you will find continual references to those concepts and the fundamental terms defined in chapter one.

Chapter two also begins the transition from fundamental electronics theories to descriptions of how satellite systems operate. Within these descriptions, the chapter compares the well-known geosynchronous satellites with other types such as low earth orbit satellites. In addition, the chapter lists frequencies used for Y. The chapter continues these discussions with an overview of uplink stations and by illustrating how downlinking functions. From there, the chapter provides information about C-band, Ku-band, and DBS services.

Manipulating Signals for Communication

Every electronic system used for communications requires a voltage source and a signal source found at the input of the device. Signals within the radio frequency band carry intelligence from the transmission point to the receiver through a process called modulation. However, because of the high frequencies used for transmission, circuits within the receiver must have the capability to drop the frequencies to a lower level and then remove the intelligence from the signal through demodulation.

Any transmission or reception of signals involves electronic circuitry that processes and manipulates the information contained within the signal. The three functions of amplification, oscillation, and switching performed by electronic stages and circuits lead to other processes such as modulation and demodulation. To varying degrees, those functions involve basic concepts such as gain, feedback, attenuation, and phase reversal.

Amplification

Amplifiers increase either the voltage, current, or power gain of an output signal. We can classify amplifiers by whether the device amplifies voltage or power. The term "amplifier" may describe a single stage or a transmitter or receiver section consisting of two or more stages of amplification. In turn, the *efficiency* of an amplifier is the ratio of signal power output to the dc power supplied to the stage by the power supply. In a multiple-stage voltage amplifier, all stages contain voltage amplifiers.

Amplifiers have constant gain over a specific range of frequencies called a band. Because this range of frequencies fits within upper and lower limits, the frequencies have a bandwidth. The upper and lower limits of the bandwidth are set by the cutoff frequencies of the amplifier. For example, an amplifier that has constant gain for a range of frequencies extending from 40 MHz to 44 MHz has a bandwidth of 4MHz. For the purposes of this example, the cutoff frequencies are 40 and 44 MHz. The specific values for the cutoff frequencies vary from amplifier to amplifier.

Although a single transistor amplifier can provide a large gain, most electronic devices require more gain than one transistor amplifier can develop. *Coupling* involves the connecting of one or more amplifier stages together through various circuit configurations. When a signal passes through two or more stages in sequence, we define the circuits as cascade stages. However, a multiple-stage amplifier designed to provide a power output will include a power amplifier in the last stage. The other stages in the amplifier will consist of voltage amplifiers.

Voltage Amplifiers

A voltage amplifier builds a weak input voltage to a higher value but supplies only small values of current. Thus, a voltage amplifier always operates with a load that requires a large signal voltage and a small operating current. A cathode-ray tube is an example of this type of load. Every voltage amplifier stage has voltage gain shown as the ratio of the output voltage to the input voltage.

Power Amplifiers

Even though voltage amplifiers increase the output voltage, the actual power output may remain at a low value. If the signal must operate some type of current-operated load, such as a speaker, the low power output will not drive the load. Power amplifiers increase the power gain of the circuit by supplying a large signal current. As the following equation shows, power gain is a product of voltage and current:

$$A_P = P_{OUT}/P_{IN} = A_V \times A_I$$

where:

A_P represents the power gain,
P_{OUT} equals the output power, and
P_{IN} equals the input power

Oscillation

Electronic systems require one or more alternating voltages or currents to reproduce sound and pictures. In a radio communications transmitter, the system applies a radio frequency voltage or current to an antenna for the purpose of producing radio waves. When considering the different types of voltages in a receiver, always remember that the alternating current used to produce the radio waves is different than the alternating current present in power lines. Radio waves have a much higher frequency than ac power lines and may require different shapes of waveforms than the sine waves seen with power lines.

As you know, 60 Hz ac line voltages produce sinusoidal waveforms. However, the proper operation of circuits within a receiver may depend on a sawtooth, triangle, or a rectangle shaped waveform. Because of this, the alternating voltages and currents must be produced by circuits within the electronic system called oscillators.

An *oscillator* consists of an amplifier with a feedback circuit connected to the output of the amplifier and converts a dc voltage to an ac signal. With the feedback circuit feeding back a portion of the output to the input of the amplifier, the amplifier begins to produce its own input. For this to occur, the feedback circuit must produce a voltage or current of the *proper phase* and *proper amplitude* to the input.

Referring to *Figure 2.1*, the amplifier has an output greater than the magnitude of the input. Moreover, the output signal has the opposite polarity of the input signal. The opposite phase, greater amplitude output signal travels through the feedback circuit where a phase reversal and a reduction in amplitude occur. As a result, the signal that feeds back to the amplifier input is identical to the input signal and works as the amplifier input.

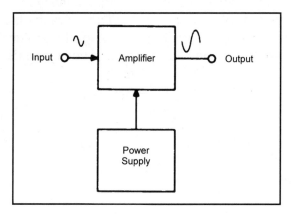

Figure 2.1. Output of an amplifier circuit.

The oscillator operation is triggered by small changes in voltage or current at the input when the oscillator turns on. Once the oscillator starts, the whole operation repeats itself to maintain the oscillations. Oscillators vary in stability or the ability to maintain an output that has a constant frequency and amplitude. Different circuit configurations and oscillator types affect stability with well-designed oscillator circuits maintaining a constant rate of oscillation over a long period of time. The loss of amplitude seen for progressive cycles of oscillation is called *damping*.

Switching

Switching circuits either generate or respond to nonlinear waveforms such as square waves and are a fundamental component of many different types of electronic systems. For example, transistors work as switches when driven between saturation and cutoff. In practice, a square wave input causes the transistor to bias off when the input signal has a specific negative level and to saturate when the input signal has a specific positive level. As a result, the transistor performs the same role as the opening and closing of a mechanical switch.

During operation of a switch, the specific negative value of the input signal equals the emitter supply connection. Regardless of whether the emitter supply equals 0, 5, or 10 volts, the negative value of the input signal for cutoff to occur equals the emitter supply value. The specific positive value of the input signal equals the supply voltage to the collector. Again, depending on the circuit requirements, the positive value of the input signal needed to cause saturation equals the collector supply voltage value.

Along with bipolar transistors, field-effect transistors, metal-oxide silicon FETs, and operational amplifiers are also effective as switches. Transistor switches are measured in terms of rise time, or the amount of time required to go from cutoff to saturation, delay time, or the time required to come out of cutoff, storage time, or the time required for a transistor to come of saturation, and fall time, or the time required for a transistor to go from saturation to cutoff. FET switches are measured in terms of turn-on time, or the sum of delay time and rise time, and turn-off time, or the sum of storage time and fall time. Operational amplifier switches are measured in terms of slew rate or the maximum rate at which the device can change its output level.

Basic switching circuit configurations include the inverter and the buffer. An inverter produces a 180-degree voltage phase shift and responds to a low input voltage signal with a high output voltage signal. In contrast, a buffer does not introduce a 180-degree phase shift. From the perspective of component configuration, a buffer is either an emitter-follower, a source-follower, or a voltage-follower.

Gain and Bandwidth vs. Attenuation

Gain expresses the ratio of input signal voltage, current, or power to the output signal voltage, current, or power of an amplifier. Gain can be measured in terms of current, voltage, or power. Because amplifiers should have predictable output values, the value of amplifier gain should remain stable under normal operating conditions.

Current gain, or A_i, occurs when the amount of ac current flowing through an amplifier from input to output increases. The equation for finding the value of current gain is:

$$A_i = i_{out} / i_{in}$$

where:

i_{out} equals the current at the output of the amplifier and
i_{in} equals the current at the input of the amplifier.

The current gain of a multistage amplifier is the product of the individual stage current gain values.

Voltage gain, or A_v, occurs when the amount of ac signal voltage increases from the input to the output of an amplifier. The equation for finding the value of voltage gain is: $A_v = v_{out}/v_{in}$, or the value of the voltage at the output of the amplifier divided by the value of the voltage at the amplifier input. The voltage gain of a multistage amplifier is the product of the individual stage voltage gain values.

Power gain, or A_p, occurs when the ac signal power increases from the input of the amplifier to the output. The equation for finding the value of power gain is: $A_p = A_i A_v$, or the value of the current gain multiplied by the value of the voltage gain. The power gain of a multistage amplifier is the product of the values of the multistage voltage gain and the multistage current gain.

Every amplifier has *bandwidth*, or a frequency range over which the amplifier has relatively constant gain. Amplifiers have constant gain over a specific range of frequencies called a band. Because this range of frequencies fits within upper and lower limits, the frequencies have a bandwidth. The upper and lower limits of the bandwidth are set by the cutoff frequencies of the amplifier. For example, an amplifier that has constant gain for a range of frequencies extending from 40 MHz to 44 MHz has a bandwidth of 4 MHz. To take this a step further, a signal covering a wide band of frequencies can carry more information than a signal covering a narrow band of frequencies.

Attenuation is the opposite of gain and is shown as loss in terms of decibels. With attenuation, the output signal from an electronic circuit has a lower amplitude or signal level than the input signal. Attenuation becomes more of a factor when high frequencies are transmitted throughout a system. The lack of attenuation exists as one of the key benefits given through the combining of the low noise amplifier and block downconverter within a single LNB package. In addition, minimizing the length of cable runs and limiting the angles of any cable bending also lessens attenuation.

Feedback

Almost every type of signal generation technique uses feedback to control the frequency, or period, and the amplitude, or level, of an output signal. Looking at *Figure 2.2*, a feedback signal travels from the output signal to the input of the amplifier. At the input, the feedback voltage modifies the control voltage that determines the size and shape of the output signal. When the feedback voltage has the same polarity or phase as the input signal, we define the feedback signal as positive feedback. Oscillators rely on positive feedback to sustain the oscillations of the amplifier. The result of positive feedback is the increase of an input signal. A device conducting current and utilizing positive feedback will conduct more current.

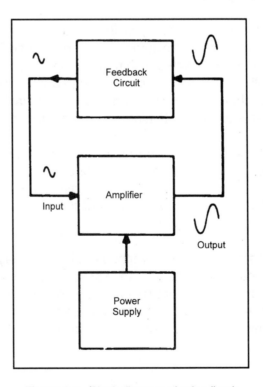

Figure 2.2. Block diagram of a feedback circuit.

Other circuits rely on negative feedback or the feeding back of part of an output signal to reduce the size and shape of the output signal. In this case, the negative feedback signal has an opposite phase or polarity as seen with the input signal. Many circuit designs use negative feedback to control the gain of an output signal from an amplifier.

Modulation

Modulation involves the encoding of a *carrier wave* with another signal or signals that represent some type of intelligence. With modulation, an audio frequency signal affects the frequency or amplitude of radio frequency waves so that the waves represent communicated information. A carrier is a single, unmodulated radio frequency signal. Either amplitude modulation or frequency modulation superimposes information onto the carrier and produces side-carrier, or sideband, frequencies that extend above and below the original frequency. As the name suggests, the carrier wave carries the transmitted signal.

When modulated, the carrier frequency is the center frequency of the modulated wave. Carrier waves are equal to the sum of the unmodulated RF carrier and two RF sideband frequencies. Signals with a high modulation frequency lie farther from the carrier than signals with low modulation frequencies. Consequently, a signal such as a broadcast television channel containing both low and high frequencies requires a greater bandwidth than a signal containing only low frequencies.

Radio waves can only carry signal information when modulated by another signal. While a perfect unmodulated carrier has zero bandwidth and contains no information, the modulated signal occupies a bandwidth least comparable to the modulating signal. Modulation combines the waveforms of the combined signals and yields different combinations of those signals.

Several different types of modulation methods exist and may be seen in various types of communications equipment. As an example, AM, or amplitude modulation, and FM, or frequency modulation, are used to transmit the picture and sound information in television systems. Along with the oft-used AM and FM methods, some communication systems use phase modulation and single-sideband modulation.

Amplitude Modulation

With amplitude modulation, the amplitude of the RF carrier shown in *Figure 2.3A* changes while its frequency remains constant. The changes in the carrier amplitude vary proportionately with the changes in the frequency of the audio frequency modulating signal shown in *Figure 2.3B*. Commercial AM radio stations operate with broadcast frequencies in the 550 to 1650 kHz range. When considering television signals, the video signal is amplitude modulated. In addition, many amateur radio operators use amplitude modulation for their signal transmissions.

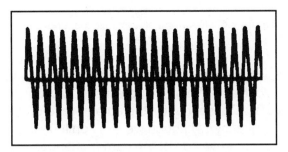

Figure 2.3A. RF carrier waveform.

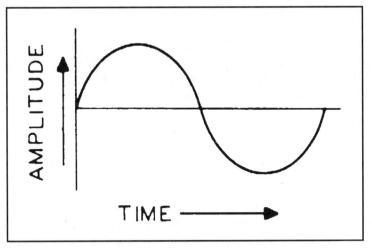

Figure 2.3B. Audio frequency modulating signal.

As mentioned earlier, the constant amplitude *carrier wave*, or *CW*, does not contain any modulating information. The thin outline drawn along the peaks of the modulated carrier wave is called the modulation envelope. Exact reproduction of the transmitted signal at the receiver requires that the modulation envelope produced at the transmitter has the same waveform as the modulating signal.

The waveform shown in *Figure 2.4* shows a amplitude modulated waveform resulting from the combination of a 1 MHz local oscillator frequency and an audio signal produced by speaking into a microphone at an RF amplifier. A modulator in the system varies the effective voltage obtained from the power supply so that the supply voltage to the RF amplifier either doubles or drops to zero on audio peaks.

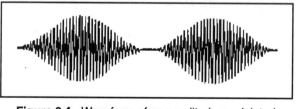

Figure 2.4. Waveform of an amplitude modulated carrier.

The doubling of the power supply voltage causes the corresponding amplitude of the output RF signal also doubles. When the power supply drops to zero, the output RF signal also drops to zero. Consequently, the output waveform found at the amplifier is a modulated radio frequency. Although we can continue to see the local oscillator frequency, the amplitude or envelope of the carrier changes at a rate determined by the modulator and the voice signal found at the microphone. Referring to *Figure 2.4*, low-level audio signals produce the smaller peaks and valleys in the signal while larger level audio signals cause larger changes to appear.

The modulated RF output signal divides into three individual components. Those are the original oscillator frequency and two sideband frequencies. The upper sideband frequency results from the adding of the modulating and oscillator frequencies while the lower sideband frequency is the difference between the modulating and oscillator frequencies.

Frequency Modulation

With frequency modulation, the frequency of the carrier changes while its amplitude remains constant. As the modulating signal increases to a maximum positive value, the carrier frequency also changes. When the modulating frequency drops to zero, the carrier frequency decreases to its original value. As the modulating frequency increases to its maximum negative value, the carrier frequency decreases.

Looking at *Figure 2.5*, the positive peak of the sine wave coincides with an increase in oscillator frequency. As a result, the change in frequency of the frequency modulated carrier wave corresponds with a change in the amplitude of the input signal. The rate of the frequency changes correspond with the modulating frequency. As opposed to amplitude modulation, the amplitude of the frequency modulated waveform in the figure remains constant.

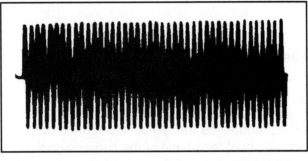

Figure 2.5. Waveform of a frequency modulated RF carrier.

As an example of frequency modulation, a microphone controls the capacitance of an oscillator. Thus, any voice fluctuations picked up by the microphone vary the spacing within the capacitor and change the oscillator resonant frequency. With this change, the desired frequency modulation is introduced into the circuit.

Figure 2.5 also shows that the carrier frequency varies above and below a center frequency. We refer to the amount of positive or negative change in the carrier frequency as *deviation*. While the amplitude of the modulating signal determines the deviation, the frequency of the modulating signal determines the rate that the carrier varies above and below the center frequency.

Phase and Single-Sideband Modulation

Both the phase and single-sideband modulation methods suppress the carrier and one sideband while transmitting one sideband from the original sideband signal. Because the upper and lower sidebands also include the modulating information, a typical AM demodulation scheme suppresses the carrier wave and one sideband and transmits the remaining sideband. This type of transmission is referred to as single sideband transmission, or SSB.

Pulse Modulation

Digital communication systems utilize another method called pulse modulation; the system converts the intelligence held within the modulating signal into a pulse. After the conversion occurs, the system pulses the RF signal for the type of pulse modulation used. The modulating pulses may control the amplitude, frequency, on-time, or phase of the carrier.

Quadrature Phase Shift Keying Modulation

With digital satellite signals, a different modulation method called Quadrature Phase Shift Keying, or QPSK, modulates the digital information onto the carrier. This occurs through the modulating of the phase of the carrier signal. During operation, the modulating circuit senses the type of data and forces the carrier into one of four different phase states called a symbol. Each symbol contains two data bits. As a result, QPSK doubles the potential amount of data transmitted through amplitude or frequency modulation.

Quadrature phase shift key modulated data rates occur as symbol rates rather than as a bit rates. As an example, a symbol rate shown as 20MS/s or 20 mega-symbols equals 40Mb/s or 40 megabits bits per second. *Figure 2.6* depicts each possible pair of data bits represented by a different phase angle while *Figure 2.7* shows a QPSK waveform.

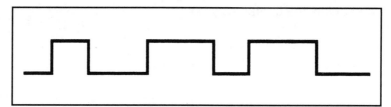

Figure 2.6. QPSK modulated data.

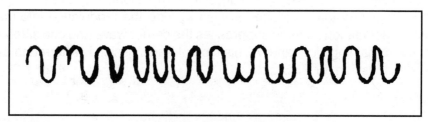

Figure 2.7. QPSK waveform.

Heterodyning

Downconverting the high frequency incoming signals to usable low frequency signals involves a process called *heterodyning*. During the process of heterodyning, the output frequencies from two signal sources combine in a mixer circuit. The resulting output signal is an algebraic combination of the two input frequencies. Many different types of receivers use heterodyning, or "beating," to combine the output of a local oscillator with an incoming RF signal for the production of the IF signal.

Referring to *Figure 2.8*, the output signal from an RF amplifier and the output signal from the local oscillator mix at the mixer input and begin the non-linear process of heterodyning. The mixer stage output signal consists of the original RF signal, the original local oscillator signal, the sum of the original signals, and the difference of the original signals. However, because the operation of the system dictates that only the difference of the two input frequencies is desired, a bandpass filter at the output of the circuit allows only the difference frequency to pass. In this case, the radio frequency of 1000 kHz heterodynes with a local oscillator at a frequency of 1,456 kHz and produces the desired difference frequency of 456 kHz.

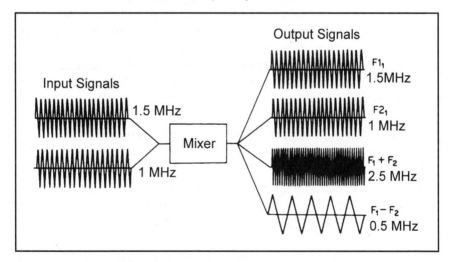

Figure 2.8. Representation of heterodyning.

Intermediate Frequencies

The heterodyning of the RF signal and local oscillator output signal produces a band of *intermediate frequencies* that have much lower values than those seen in the radio frequency band. Each frequency in the intermediate frequency band equals the difference between the oscillator frequency and the frequency of a corresponding wave in the RF channel. Because the center frequency of the received RF carrier also heterodynes with the local oscillator output, it produces the center frequency for the intermediate frequency band. The center frequency of the intermediate frequency band is called the *IF carrier*.

Superheterodyne Receiver Operation

Many different types of electromagnetic waves induce voltages and currents at the antenna of a radio receiver. To ensure proper operation, a receiver must have the characteristic of selectivity, or the ability to select one specific frequency while rejecting all others. In addition, a receiver must have sensitivity, or the ability to adequately amplify a weak desired frequency or RF carrier.

When we consider the basic ingredients needed for the reception of an RF signal, several building blocks fall into place. First, the receiver requires some method, such as a tuned circuit, for selecting the desired frequency. Second, the receiver must have a method for amplifying the desired frequency and then for converting the RF signal into a lower, easily-handled intermediate frequency. The conversion of the RF signal into an IF signal reduces the need for multiple tuned circuits. Third, the receiver requires a method for removing the intelligence from the modulated signal.

Figure 2.9 uses a block diagram to depict the operation of a superheterodyne receiver. Receivers that operate with a fixed intermediate frequency are called *superheterodyne receivers*. Used in all radio, television, and satellite receivers, the superheterodyne design provides high selectivity, sensitivity, gain, and reliability because the circuits operate at the relatively low intermediate frequencies.

Referring to *Figure 2.9*, tuned circuits select the desired frequency while the RF amplifier amplifies the RF carrier and the carrier sidebands. Then, the incoming RF signal heterodynes with the frequency of the local oscillator contained within the receiver. The mixer stages converts the RF signal to the IF signal.

After heterodyning occurs, the intermediate frequencies go through several stages of IF amplification. Because the IF carrier contains the same amplitude or frequency modulation as the received carrier wave, the detector removes the intelligence contained by the modulating signal from the IF carrier. Superheterodyne receivers employ several stages of amplification between the mixer and the detector.

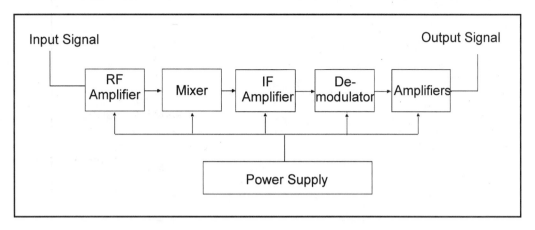

Figure 2.9. Block diagram of a superheterodyne receiver.

Demodulation

Also referred to as *detectors*, demodulators decode and recover the intelligence from the carrier signal. When considering the demodulation of amplitude modulated signals, the demodulator circuits detect the envelope that corresponds with the modulating signal and eliminate the carrier wave. In the simple diode detector circuit depicted in *Figure 2.10*, the modulated RF signal travels from output of the IF stages and encounters the detector. The diode detector allows only half of the modulated RF waveform to pass. Because the filter capacitor cannot follow the RF signal, it passes only the information-carrying envelope of the waveform.

The detection of a frequency modulated signal for the recovery of audio intelligence is more difficult than the detection of an amplitude modulated signal. Because of the characteristics of frequency modulation, a circuit called a frequency discriminator combines with a limiting circuit to ensure that the amplitude of the signals remains constant. In those circuits, changes in frequency cause a change in the output voltage from the sound carrier. Other demodulator circuits decode and recover color information contained in sidebands from the subcarrier. Because the color subcarrier is phase-modulated, the demodulator uses the phase relationship of the color difference signals to decode the modulated signals.

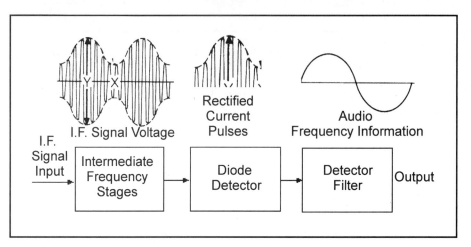

Figure 2.10. Block diagram of a simple diode detector.

Antennas

Figure 2.11 shows a simple dipole antenna. As the figure shows, the dipole consists of two pieces of wire cut to a specific length. In this case, the figure represents a half-wave dipole antenna. Because of the configuration, the antenna has a length of about 2.5 feet and resonates at channel 11 VHF signals that occur at 199 MHz. As resonance occurs, the antenna becomes receptive to the signals. The signal received at the antenna is a signal voltage. During transmission and reception, the signal voltage transfers from a power amplifier in the transmitter; through the air; and to the dipole antenna. As the voltage transfers to the antenna, the antenna element acts as a capacitor and charges and discharges. An electrostatic field that varies in strength and frequency exists between the two rods that make up the antenna and creates an electromagnetic field. If used for a television, variations in the electromagnetic field produce pictures and sounds at the receiver.

Using Satellites to Communicate

Today, satellite communications have become commonplace to the point that many consumers simply expect the communications to occur. Satellite communications involves the relaying of electromagnetic waves that carry some form of information from one or more uplinks to a satellite and then to an unlimited number of receiving stations on the earth. The information carried on the waves may consist of video, audio, or data signals.

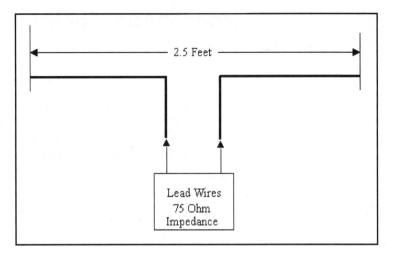

Figure 2.11. Diagram of a simple dipole antenna.

As all of us know, the reception of television signals through satellite technologies has revolutionized the way that we communicate. In 1962, the United States launched TELSTAR, the world's first satellite used for television programming. Because the methods for launching satellites still had to evolve, TELSTAR operated in an elliptical orbit that did not allow the satellite to remain stationary over one area. As a result, programming through TELSTAR only occurred a few hours each day. The first satellite to park in a geostationary orbit, Early Bird, flew in 1965 and relayed telephone calls, data, news, and television programming along the Eastern seaboard.

With the steady improvement of satellite and broadcast technologies, a single satellite placed in a geostationary orbit can replace thousands of local broadcast facilities and bring a myriad of programming choices directly to an individual's home. With this, the chains of distance have suddenly broken. Given the technology, we can watch sporting events from around the world or gain near-instant insight into political and societal happenings. The broad distribution of satellite television programming began during the mid- to late-1970's as corporations such as HBO and WTBS found that the technology allowed effective delivery of their products to a wide audience. Today, individual C-band, Ku-band, and direct broadcasting system earth stations that have the combined capability to receive more than 200 programming channels dot the globe.

Satellites in Geosynchronous, Inclined, and Low-Earth Orbits

Some of the most familiar communications satellites lie within geosynchronous orbits. Others use inclined orbits while still others utilize low-earth orbits. Each type of satellite orbit yields different results and requires different types of receiving equipment. As an example, most satellites used for television programming rely on geosynchronous orbits. Newer satellites used for cellular phone and pager communication have low earth orbits.

Geosynchronous Orbits

During 1945, the scientist and science fiction writer Arthur C. Clarke spoke about the placement of an artificial satellite in earth orbit at 35,803 kilometers above the equator. In this geosynchronous orbit, a satellite will orbit the earth at the same speed as the earth's rotation and remain stationary with respect to any point on the earth's surface. Because of Clarke's prediction, we now refer to the equatorial belt as the Clarke Belt.

Satellites placed orbiting within the Clarke Belt take a place called an orbital slot. That is, each satellite occupies a specific slot or subdivision. During transmission, satellite uplink stations transmit signals to the satellite. In turn, the electronic circuitry on the satellite processes the signals and then transmits the signals through a downlink to a receiving station on the earth.

Inclined Orbits

We define the measure of inclination as any difference between the angle made by the plane of the spacecraft with respect to the plane of the equator. Geosynchronous satellites have a 0.1° inclination. However, satellites in an inclined orbit have inclinations that range up to three degrees. Rather than remain stationary in its assigned slot, a satellite placed in an inclined orbit follows a figure-eight pattern about its assigned location. The width of the figure eight increases as the inclination and north/south motion increase.

Because of this, a satellite reception dish receiving signals from an inclined orbit satellite must have the ability to track the satellite. This tracking ability occurs through adjustments in the polar mount elevation or the azimuth. Tracking outside a given range occurs through adjustments in both the elevation and the azimuth.

Low-Earth Orbit Satellites

A low-earth orbit, or LEO, satellite circles the earth at an altitude of only 200 to 500 miles. Because of the proximity to the earth's gravitational field, LEO satellites travel at 17,000 miles per hour and complete one orbit within 90 minutes. Depicted in *Figures 2.12*, some LEO satellites operate as remote sensing and weather satellites because of the capability to capture highly detailed images of the earth's surface from low altitudes.

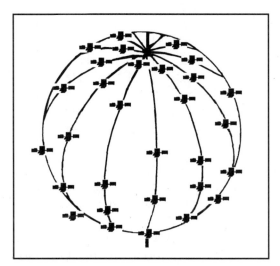

Figure 2.12. Representation of low-earth orbit
satellite orbital paths.

In addition, low-earth orbit satellites also serve one of the hottest trends in satellite technology by supporting global personal communications networks. As an example, Iridium™ by Motorola exists as the world's first handheld global satellite telephone and paging network with 66 LEO satellites. The service offers global communications access through cellular phones or pagers. Other telecommunications providers such as Teledesic, Globalstar, ICO, Ellipso, Astrolink, and Spaceway rely on low-earth satellites for not only cellular and pager communications but also worldwide, high speed access to computer networking, broadband Internet access, high-quality voice, and other digital data services.

Transmitting Signals from Earth Stations to Satellites and to Your Receiver

The transmissions to and from the satellite occur at microwave frequencies that exist within the gigahertz range. As shown in *Chart 2.1*, microwave signals occupy a frequency range much higher than that used for traditional television signals. Because microwave signals occur at high frequencies, those signals can carry large quantities of information. The bandwidth of the signal increases as the frequency increases.

L-Band	1-2 GHz
S-Band	2-4 GHz
C-Band	4-8 GHz
X-Band	8-12 GHz
Ku-Band	12-18 GHz
K-Band	18-27 GHz
Ka-Band	27-40 GHz
V-Band	40-75 GHz
W-Band	75-110 GHz
mm-Band	110-220 GHz

Chart 2.1. Microwave frequency bands.

In addition to having attractive bandwidth characteristics, microwave signals also exhibit less signal degradation due to noise during uplinking than lower frequency signals. Furthermore, the high frequency microwave signals used to carry satellite communication pass through the upper atmosphere easily. RF signals at frequencies lower than 30 MHz reflect off the ionosphere back to earth.

Uplinking the Satellite Signal

A satellite uplink station transmits the signals to the satellite at different frequencies because of the need to avoid interference with the downlink signals. All signals transmitted from the uplink station occur at high frequencies in the 1 to 40 GHz range and rely on an extremely narrow transmission beam. Without the higher frequencies and narrow beam, the transmitted signals would interfere with both the downlink signal from the satellite and signals at other satellites located within the arc. The uplink station transmits control signals for the satellite. In part, these signals ensure that the satellite retains the correct orientation and orbit. Along with this, the signals control the accuracy of the signal reproduction and timing.

Downlinking the Signals

Whether used for the transmission of video, audio, or data signals, every satellite has a number of transponders, or sets of physically separate onboard communications circuitry. Each transponder accesses a pair of receive/transmit antennas and associated electronic circuits for a designated channel or set of channels. Once the uplink station sends signals to the satellite, the circuitry converts the high uplink frequencies to a lower frequency signal and then uses high power amplifiers to retransmit the signals back to earth.

Receiving the Signal

Figure 2.13 shows a simple diagram of a television receive only system. As the figure shows, satellite television receiving stations use a dish shaped antenna to collect and focus the weak, high frequency signal towards a feedhorn. Because the microwave signals transmitted from the satellite have an extremely short wavelength and low power, obstructions such as trees or, in some cases, heavy rainstorms can block the signals. In addition, background noise such as terrestrial interference can also ruin the signal.

Within the feedhorn, a small antenna absorbs the captured signal. Circuitry attached to the antenna amplifies and converts the signal to a lower frequency. From there, the lower frequency signal moves to a receiver for processing, additional down conversion, and further amplification.

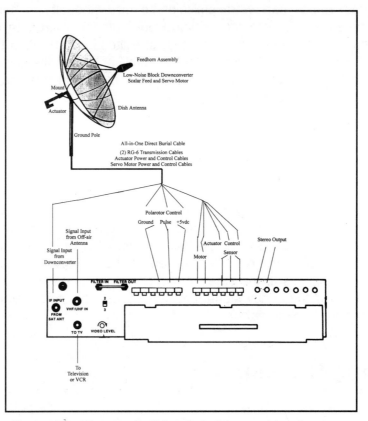

Figure 2.13. Diagram of a C-band television receive-only system.

C, Ku, and DBS Signals

As satellite television reception has evolved, viewers have had a range of options. During the early days of satellite television technology, large C-band dishes served as the standard approach for receiving signals. Later, Ku-band dish and receiver technologies provided promise of small dish reception. However, large scale Ku-band transmissions for the consumer never materialized and few Ku-band-only systems existed. The advent of direct broadcast systems that featured a large variety of programming choices along with small footprint dishes generated new life for the satellite television reception market.

C-Band Satellite Signals

C-band satellite TVRO systems receive microwave signals in the 3.7 to 4.2 GHz range and provide benefits not seen with the DBS systems. Some of those benefits are:

- The ability to select satellites from across the band. Any C-band system can move back and forth across the satellite band through the use of an actuator motor.

- Expansive programming choices. Because the dish can move across the band, the system is not limited to the programming choices found on one satellite.

- The capability to pick up special feeds

- Reliability in all types of weather

- The capability to receive Ku-band and 4DTV transmissions

Given these benefits, C-Band systems remain as a viable alternative. As those systems age, routine maintenance may require the replacement of the polarotor assembly, the upgrade of the LNB to a lower temperature unit, the replacement of cables or cable connectors, or the replacement of actuator motors. Along with these possibilities, service may also include realigning the dish or installing terrestrial interference filters.

In addition, some may also decide to replace their current receivers with a model that offers more features for a lower price. As an example, many popular receivers offer basic features such as favorite video channel memory, satellite position memory, C/Ku-band capability, onscreen display; parental lockout; picture-in-picture; and a VCRS decoder module.

In addition to those features, high-end receivers offer high-speed data ports, the capability to receive MPEG-2 audio signals, an interactive program guide, instant pay-per-view, and program timers. Because the receiver works with digital as well as analog signals, it also offers compatibility with the HDTV transmissions that will become a standard part of the television broadcast world.

Ku-Band Satellite Signals

Ku-band TVRO systems receive microwave signals in the 10.95 to 12.75 GHz range rather than the 4 GHz signals seen with C-band systems. In contrast to the 5 to 8.5 watts of transmission power seen with C-band systems, Ku-band transmissions have a signal power of approximately 25 to 45 watts. Given the higher frequencies and higher power of the Ku-band transmissions, we can use smaller dishes to receive the signals. Moreover, the Ku-band signals do not have the susceptibility to terrestrial interference that can occur with C-band systems.

Direct Broadcast Satellite Signals

A direct broadcast system uses the same principles seen with the C- and Ku-band systems in that many channels of programming are delivered directly from the satellite to a dish and receiver located at an individual house or business. Direct broadcast satellite signals are broadcast in the microwave Ku band, have a frequency of 12.5 GHz, and utilize encoded digital signals. *Figure 2.14* shows a typical direct broadcast system manufactured by Thomson Electronics/RCA. The antenna has a gain of 34 decibels and a half-power beamwidth of 3.5 degrees.

Figure 2.14. Direct broadcast
system manufactured by
Thomson Electronics.

Each DBS satellite has 16 transponders that operate at 40 megabits per second. Because direct broadcast systems utilize compressed digital signals rather than analog signals, a DBS system does not move from satellite to satellite. Instead, the 18" dish is fixed to one satellite location. The use of digital signals allows each transponder to send several channels. As a result, the owner of a DBS system may choose from 120 - 200 channels of popular programming.

Types of Available Programming

A relationship exists between the design of the satellite and the type of signals accepted from the uplink station and then transmitted to the receiving stations. Satellites launched during the early 1970s could carry twelve television programs on signals using linear polarity. Given the 3.7 to 4.2 GHz frequency range of the C-band signals, the 500 MHz divided into twelve segments with each segment having a width needed for the transmission of the high quality video and audio information needed for television pictures. The twelve channels had a bandwidth of 36 MHz and 4 MHz guard bands that eliminated any chance of crosstalk between channels. Individual transponders on the satellite matched with each channel.

The use of the 36 MHz bandwidth allowed the transmission of video signals within 28 MHz of the band and audio and/or data signals within the remaining portion of the band. Generally, this practice continues with analog satellite broadcasts. The use of digital transmission methods for satellite television broadcasts has effectively changed this by compressing more information into existing bandwidth and by introducing higher bandwidth technologies. In addition, some broadcasters will reduce the bandwidth available for video signals, and while accepting some degradation, increase the carrying capacity of the satellite transponder.

As satellite technology began to mature, newer satellite designs doubled the number of channels transmitted within the 500 MHz bandwidth. The capability to provide the additional channels occurred through the transmission of all odd channels with vertical polarity and all even channels with horizontal polarity. With the earth station receivers handling only one polarity at a time, frequency overlap could exist without causing interference between the opposite polarity channels.

Referring to *Chart 2.2*, the standard method for transmitting C-band signals involves the use of the 500 MHz bandwidth. The center frequency within the 500 MHz bandwidth serves as the downlink frequency. In most cases, domestic and international satellite providers use this format for their designs. However, special needs sometimes dictate the use of other combinations of center frequencies and channel bandwidth.

Transponder	Downlink Frequency in Megahertz	Frequency Band in Megahertz
1	3720	3702 - 3738
2	3740	3722 - 3758
3	3760	3742 - 3778
4	3780	3762 - 3798
5	3800	3782 - 3818
6	3820	3802 - 3838
7	3840	3822 - 3858
8	3860	3842 - 3878
9	3880	3862 - 3898
10	3900	3882 - 4008
11	3920	3902 - 3938
12	3940	3922 - 3958
13	3960	3922 - 3958
14	3980	3962 - 3998
15	4000	3982 - 4018
16	4020	4002 - 4038
17	4040	4022 - 4058
18	4060	4042 - 4078
19	4080	4062 - 4098
20	4100	4082 - 4118
21	4120	4102 - 4138
22	4140	4122 - 4158
23	4160	4142 - 4178
24	4180	4162 - 4198

Chart 2.2. C-band channel listing.

Video Signals

Video signals consist of the sync, luminance, and chrominance information recorded by a video camera, processed at the station site, and then uplinked to the satellite. While the sync pulses have a consistent amplitude and spacing, the changing amplitude and spacing of the luminance and chrominance signals represent the changes occurring in a transmitted picture. By definition, luminance signals represent the amount of light intensity given by a televised object, cover the full video-frequency bandwidth of 4 MHz, and provide the maximum horizontal detail.

Television broadcast standards place each channel within a 6 MHz band. Within each 6 MHz band, the picture information carrier wave lies 4.5 MHz below the sound information carrier wave. *Figure 2.15* shows the separation of the signals on a frequency response curve. The desired picture signal, which includes both sidebands and sync signals, covers 4 MHz of the bandwidth while the desired sound carrier and its sidebands cover only 50 KHz.

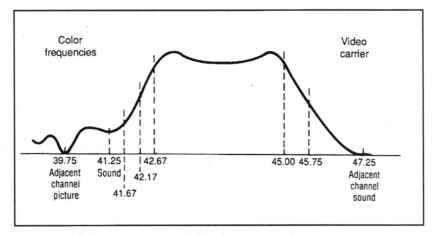

Figure 2.15. Television signal frequency response curve.

Audio Signals

With traditional television signals, the audio signal arrived as a subcarrier attached to the video signal. Satellite reception systems either carried the signals as subcarriers or assigned non-video transponders to carry the signals in the SCPC, or single-channel-per-carrier mode. With this, a wide-range of audio entertainment becomes available to the system owner. For example, most major league baseball teams have satellite radio networks. Moreover, all the radio networks provide their affiliates with multiple newscasts per hour. State radio networks, talking book networks for the blind and special feeds also exist. With the entrance of digital video, audio and data signals into the marketplace, newer systems broadcast the audio signals as an integrated portion of the digital video/audio/data transmission.

Receiving SCPC Signals

The SCPC mode allows the broadcasting of audio messages separate from video carriers. During an SCPC transmission, the transmitting station sends each audio channel on its own carrier. As a result, the SCPC carrier and the voice-grade messages occupy only a small portion of the entire bandwidth. Given its large bandwidth carrying capacity, a satellite transponder can carry hundreds of SCPC audio, data, or telephone signals that have individually assigned carriers and frequencies. A specially designed FM radio receives the SCPC signals.

The Internet

Many businesses and consumers use a derivative of the direct broadcast satellite systems to receive the Internet. The technology allows a computer user to download information from the Internet at 400 kilobytes per second per minute. A typical DirecPC system includes the satellite receiving system and an internal card for the personal computer. During operation, the user dials into the Internet through a local Internet Service Provider and a standard modem. Any requests for data or file transfers travel to the server via the dial-up modem. Once this occurs, the Internet server sends the data to a network operations center for uplinking to the satellite. After processing occurs, the satellite transponder downlinks the data to the personal computer at high transfer speeds.

Specialized Data

As mentioned, several corporations have begun utilizing low-earth orbit satellites for the transmission of business data, messaging services, and two-way monitoring. Applications for those data transmissions include the monitoring of fixed assets such as electric utility meters, oil and gas storage tanks, wells and pipelines, and environmental projects. In addition, the low-earth orbit satellite technology has become especially useful for the tracking of mobile assets such as commercial vehicles, trailers, rail cars, heavy equipment, and fishing vessels. The communications services offered through satellite networks of this type introduce communications to locations underserved by traditional telecommunications industries.

Terrestrial Interference

Any type of unwanted earth-based communication signal falls within the category of *terrestrial interference* and causes different types of problems for a satellite receiving system. Terrestrial interference, or TI, includes high frequency microwave signals used for point-to-point transmissions to lower frequency audio signals, and video signals. As you may suspect, we can separate TI into signals that range from 1 GHz to 8 GHz and signals that range below 1 GHz. The higher frequency signals enter the system through the antenna or low noise amplifier while the lower frequency signals enter through connectors or improperly grounded transmission cables.

The symptoms associated with terrestrial interference range from sparklies—a confetti-like colored interference found throughout the reproduced picture—to the complete whiting out of the picture. Interference in the 5 MHz to 1 GHz range causes symptoms such as:

*   Co-channel interference where a second channel bleeds into the desired channel or causes a venetian blind effect to appear in the reproduced picture

*   Herringbone patterns that appear in the reproduced picture, and

*   Audio buzz

With low frequency interference, any of the symptoms could appear on all channels.

Interference in the high frequency range can cause a variety of symptoms to appear on all channels or just channels that receive the same type of signal format. As an example, a signal received from a microwave transmitter used by a telephone company could cause the deterioration of the lower channels found at the satellite receiver. The same problem could also cause the loss of only the low channels or only the high channels. Other symptoms of high frequency terrestrial interference include sparklies or the complete whiting out of the picture.

Sparklies

When working with satellite systems, we often hear the term "sparklies" as a reference to noise. Sparklies consist of impulse noise and appear as long duration, colored or black-and-white dots that have a trailing edge. Within the audio signal, the impulse noise takes the form of hiss or a crackling sound. The presence of sparklies tells us that the signal reaching the demodulator has too low of a level. When the signal level reaches a minimum level, the receiver can no longer discern between the desired signal and noise—and begins producing its own noise.

Summary

Chapter two enlarged the fundamental information first described in chapter one by moving to circuit functions and circuit operations. Circuit functions were discussed as you learned that analog and digital circuits perform the basic functions of amplification, oscillation, and switching. Circuit operations were defined along with the processes of modulation, heterodyning, and demodulation.

Each of those concepts has a direct bearing on the operation of satellite systems. The chapter reinforces that link by moving from circuit functions and operations to a detailed analysis of different satellite technologies that rely on geosynchronous, inclined, and low earth orbit satellites.   As technology improves, satellites have become useful tools for not only transmitting television signals but also cellular phone, pager, computer data, and Internet traffic.

CHAPTER

3

Dish Antennas

Chapters 1 and 2 presented information about that principles that make the transmission of data through satellite technologies possible and the components that make up a satellite system. Within those, we learned about signal bands, classifications of noise, the difference between analog and digital signals, and properties such as amplification, oscillation, and feedback. In addition, we discovered that different types of satellites exist and that those satellite systems carry varying types of information.

Chapter 3 adds to this knowledge by defining how dish antennas capture communications signals. Within those discussions, the chapter introduces terms such as azimuth, elevation, focal length, and polar axis. While introducing those terms and concepts, the chapter also presents detailed information about the design of dish antennas. The chapter concludes with discussions about finding the best location for a dish antenna, choosing the best reflector size, and combatting terrestrial interference.

The Dish Antenna

Any satellite receiving antenna, or dish, catches the signal transmitted by the satellite and reflects the signal back to a common point called the focal point. Effectively, the dish takes a weak signal that has spread over a wide area, concentrates the signal, and makes the signal stronger. Whether the dish construction consists of solid aluminum, fiberglass, or perforated mesh aluminum, the performance of the dish depends on the accuracy of the its curvature. Dish antennas may arrive as single piece assemblies or as several petals that fasten together.

The azimuth and elevation of the dish must match the azimuth and elevation of the communications satellite transmitting the signals back to earth. *Azimuth* represents the angle of the reflector from true north as the dish swings from side to side and targets a satellite. As *Figure 3.1A* shows, we can measure the azimuth in degrees from East to West. If viewing directions with the aid of a compass, North has an angle of 0°, East has an angle of 90°, South has an angle of 180°, and West has an angle of 360°. When the reflector moves East to West in the Northern Hemisphere, it covers an azimuth between 90° and 270°.

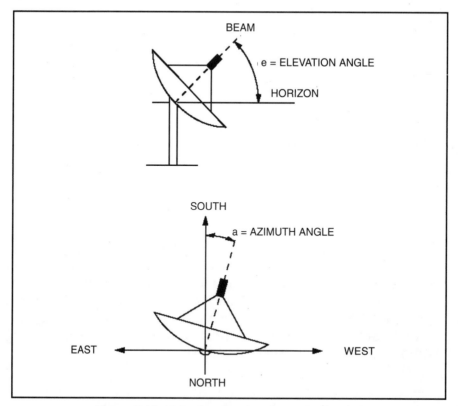

Figure 3.1A. Adjusting the elevation and azimuth.

Mathematically, azimuth, or AZ, is represented as:

$$AZ = 180 + \tan^{-1} [\tan(B) / \sin(A)] \text{ (degrees)}$$

where:

A = the latitude of the TVRO station, and
B = the longitude East of the Earth station minus the longitude East of the satellite;

We can find the degrees of longitude and minutes of latitude in any atlas. When using the equations, we must also convert the minutes and degrees into decimal degrees. With latitude, we divide the number of minutes by 60, multiply the result by 100, and add to the whole degree portion. For longitude, we subtract the number of degrees from 360°.

Also measured in degrees, *elevation* represents the upward angle that the dish must have when directed towards the satellite and is measured with respect to ground. In mathematical terms, the elevation, or EL, appears as:

$$EL = \tan^{-1}[\, m\cos(A)\cos(B)\text{-}1 \, / \, m\sqrt{1\text{-}\cos2(A)\cos2(B)}] \text{ (degrees)}$$

where:

m = 6.61 or the ratio of the geostationary orbital radius to that of the equatorial radius of the earth
A = the latitude of the TVRO station
B = the longitude of the TVRO station

The *polar axis* of a dish antenna lies parallel with an imaginary line passing through the North and South geographic poles and matches the latitude of the installation site. In addition, every dish antenna has a *focal point* where the incoming signals from the satellite focus after reflecting from the dish surface. The *focal length* of a dish measures the distance from the dish surface to the focal point.

Reflector Gain, Efficiency, and Beamwidth

A variety of factors affect the performance of a dish antenna. Those factors range from the gain given by the antenna to its beamwidth and efficiency. In turn, the design of the dish antenna and the type of materials used for the construction of the antenna are among characteristics that determine factors such as gain and efficiency.

Dish Antenna Gain

The performance of any dish antenna depends on its *gain*, or the amount of amplification that a dish provides for the received signal. By definition, gain expresses the ratio of input signal voltage, current, or power to the output signal voltage, current, or power of an amplifier. Any type of gain measurement occurs in decibels, or dBs. As an example of how gain measurements work, an amplifier that provides 3 dBs of gain doubles the signal power while an amplifier providing 10 dBs of gain magnifies the power ten times. Thirty decibels of gain amplify the signal 1000 times while 50 dBs of gain multiplies the signal by 100,000.

Antenna gain results from the size and curvature of the dish antenna, occurs by maximizing the signal at the focal point, and depends on the attenuation of any adjacent antenna signals or noise to a minimum level. A parabolic antenna focuses RF signals to the focal point in the same way that an optical parabolic lens in an optical telescope focuses light. As a result, the theoretical size of the gain of a parabola has the relationship with wavelength shown in the following equation:

$$Gain = h(pD/l)^2$$

where:

h = efficiency
D = the diameter of the reflector in meters, and
l = the wavelength of the received signal in meters

In practice, we always express gain in decibels and replace the wavelength with the frequency of the signal. By doing so, we arrive at the following equation:

$$Gain (dB) = 10 \log h + 20 \log D + 20 \log f + 20.4$$

where:

h = efficiency
D = the diameter of the reflector in meters, and
f = the operating frequency in GHz

When we examine those factors, we find that the usable size and accuracy of the reflective surface affect the antenna dish gain. Dish antenna gain also increases or decreases according to the type of signal received. As signal frequency increases, gain increases with the square of the signal frequency.

Antenna gain increases with antenna size and the efficiency of the antenna. Mathematically, antenna gain appears as:

$$G_a = 10 \log [(pd)^2 p / 100l^2] \text{ (dB)}$$

where:

p = 3.14159,
d = the diameter of the antenna,
p = the antenna efficiency as a percentage, and
l = wavelength

The gain of the antenna combines with the gain provided by the amplifier circuitry found at the antenna feed to produce a usable signal at the receiver. During the early years of home satellite systems when the amplifiers provided little gain, consumers opted for larger dish antennas. Today, high gain amplifier electronics couple with the ability to receive higher frequency signals and allow the use of smaller dish sizes.

Engineers use a standard called the *isotropic antenna* as a reference for measuring the gain of dish antennas. The isotropic standard applies an imaginary aerial antenna design that receives signals equally from all directions and has a gain of zero decibels. Companies demonstrate the capability of antennas used in real installations by showing how much more gain the antenna provides than that seen with the isotropic standard. Each of the three factors that affect the gain of the dish antenna—reflector size, accuracy of the reflective surface, and type of signal received—affect the design. *Table 3.1* provides a listing of typical gain figures for dish antenna sizes.

Dish Size	Gain
6 ft (1.8 meters)	35 dB
8 ft (2.4 meters)	37 dB
10 ft (3.0 meters)	39 dB
12 ft (3.7 meters)	40.5 dB
16 ft (5.0 meters)	43 dB
20 ft (6.0 meters)	45 dB
30 ft (9 meters)	48.5 dB

Table 3.1. Dish antenna size and gain.

Carrier-to-Noise Ratio

Every reflector antenna has a carrier-to-noise ratio and a signal-to-noise ratio. A carrier is the incoming satellite signal plus the modulation envelope that surrounds the signal. The carrier-to-noise ratio measures the radiated power from the satellite at the location and affects the system before the demodulation of the downlinked signal. We can calculate the carrier-to-noise ratio through the following equation:

$$C/N = EIRP - Path\ Loss + G/T - 10LOG(B)$$

where:

EIRP = the Effective Isotropic Radiated Power from the satellite
Path Loss = the signal loss between the satellite and the receiving antenna
G/T = the ratio of antenna gain to system noise temperature, and
B = the receiver bandwidth

All satellite receivers maximize the strength of the carrier while simultaneously limiting the amount of noise contributed by external sources or generated within the electronic circuitry that makes up the receiver. Receivers also have a threshold point expressed in decibels at specific carrier-to-noise ratios. When the carrier-to-noise ratio falls below the threshold point, noise enters the signal and causes sparklies to appear on within the picture. As the noise increases, the amount of sparklies also increase and eventually block the desired signal.

Receivers that have lower threshold points remain less susceptible to noise even when operating under low signal conditions. Unfortunately, not all receiver manufacturers measure the threshold point accurately or consistently. Therefore, evaluation of the receiver performance usually must occur with the receiver attached to a dish that has the same diameter size as the one selected for your site.

Signal-to-Noise Ratio

The signal-to-noise ratio measures variations in desired signals that occur because of the effects of noise and affects the system after the demodulation of the downlinked signal. As a result, the signal-to-noise ratio depends on the value of the carrier-to-noise ratio and the modulation characteristics of the system. Other factors that affect the signal-to-noise ratio include the de-emphasis of the receiver demodulator output and the noise weighting factor.

Preemphasis and Deemphasis

Frequency modulated systems use preemphasis, or the boosting of high frequencies in the signal, to compensate for differences in amplitude between high and low frequencies. At the point of transmission, a sound signal is a frequency-modulated signal having maximum deviation of +-25 kHz and capable of providing an audio bandwidth of 50 to 15,000 Hz. The sound carrier is transmitted at a frequency 4.5 MHz above the RF picture carrier. When considering speech, music, or any reproduced sound, the higher frequency components have less amplitude than the lower frequency components.

As a result, the signal-to-noise ratio for any frequency is lower for a high frequency audio voltage and higher for a low frequency audio voltage if the noise level remains constant. If the noise level increases at higher frequencies, then an even lower signal-to-noise ratio results. Given this characteristic, the quality of any high frequency audio reproduction suffers unless the signal-to-noise ratio improves.

At the receiver end, a deemphasis circuit compensates for the pre-emphasis applied to the transmitted sound signal. The removal of the nonlinear component from the signal allows the audio output section to reproduce a natural sound. *Deemphasis* occurs after the modulation of the sound IF signal, reduces the higher frequency noise voltages, and establishes a desired signal-to-noise ratio. With all this, deemphasis restores a balance between the higher and lower frequencies contained in any sound transmission. As a combination, preemphasis and deemphasis improve the signal-to-noise ratio of the reproduced sound.

Noise Weighting Factor

Any time that a system changes a high bandwidth signal into a lower baseband value, the signal-to-noise ratio increases. When calculating the signal-to-noise ratio, the equation will always feature standardized noise weighting figures. In mathematical terms, the signal-to-noise ratio appears as:

$$S/N = C/N + 10 \log \left[3(f_{(p\text{-}p)} / f_v)^2 \right] + 10 \log (b / 2f_v) + k_w \text{ (dB)}$$

where:

S/N = the peak-to-peak luminance amplitude of the signal to noise ratio in decibels
C/N = carrier-to-noise ratio in decibels
f(p-p) = the peak-to-peak deviation by the video signal in hertz

f_v = the highest video frequency in hertz
B = the bandwidth of the radio frequency in hertz
k_w = the de-emphasis in an FM system in decibels

Antenna Noise Temperature

Because of the wavelike properties of microwave signals, a dish antenna has the capability to capture signals from not only the side but also the back of the dish. Microwave signal energy strikes the reflective surface of the antenna and bends either because of diffraction or because of interference from other energy sources. As we know, the size and location of the side lobes shows if the dish antenna has the capability to reject the interference.

A reflector collects noise from the ground, the atmosphere, and extraterrestrial sources as well as signals. The *antenna noise temperature* rating indicates how much noise the dish antenna will detect from sources other than the desired satellite. Noise may occur because of microwave emissions from earth coming from fluorescent lights or microwave radiation. The latter type of noise can increase as the dish antenna points towards lower elevations.

The noise temperature of a reflector varies with the elevation angle, size of the reflector, frequency, and weather conditions. The amount of noise varies with dish size through the increase in temperature caused by the sidelobes of the antenna. Clear sky conditions present lower antenna temperatures than heavily clouded conditions or rain. *Table 3.2* shows the antenna noise temperatures and gain properties for different reflector sizes.

Generally, antenna noise will become a major part of the entire system noise factor unless the system relies on an LNB that has a very low noise rating. As an example, an LNB with a one-decibel noise figure will require some attention with regard to antenna noise. Prime focus dishes will have higher antenna noise temperatures than offset-fed dishes or dual reflector dishes. This occurs because of the relatively high spillover of the feed reception pattern. Because of the spillover seen with offset fed dishes, the feed will look at the high temperature ground rather than the low temperature sky.

In equation form, the total system noise temperature appears as:

$$T_{SYS} = T_{LNB} + (1 - s)T_C + sT_A \ (K)$$

where:

T_{SYS} = total system noise temperature,
T_A = effective antenna noise temperature for clear sky conditions,
T_{LNB} = the equivalent noise temperature of the LNB in degrees Kelvin
T_C = the physical temperature of the waveguide component
s   = the amount of energy passing through a medium and to the other side

Reflector Size (feet)	Antenna Noise Temperature (degrees Kelvin at 30° Elevation)	Total Reflector Gain	Increase of Antenna Noise Temperature Due to First Side Lobe (degrees Kelvin)
4	70	32.5	7.5
6	58	36	5
8	48	38.5	3.7
10	40	40.5	3,3
12	32	42.1	2.5
16	21	44..8	1.9

Table 3.2. C-band antenna noise temperatures and gain properties per reflector size.

Figure of Merit

The *figure of merit,* or *G/T,* for a reflector defines a ratio between net antenna gain and the total system noise temperature. When we consider the figure of merit for a reflector, we also consider the maximum obtainable figure of merit for a given antenna elevation angle. The usable figure of merit accounts for operational losses that occur because of antenna pointing errors, the aging of components, and increases in system noise. Reflector antennas that have a high ratio of gain to noise temperature provide much better performance.

Mathematically, the usable figure of merit appears as:

$$G/T_{usable} = 10 \log [10^{0.1(G + a + b)} / T_{SYSrain}] \ (dB/K)$$

where:

G = the gain of the antenna in decibels

a = the coupling loss given by waveguide components shown in decibels

b = losses that occur due to antenna pointing errors; the aging of components, and polarization errors

$T_{SYSrain}$ = the total system noise temperature plus increases caused by precipitation

Dish Antenna Efficiency

Each of the factors listed during our discussion of dish antenna gain also affects the *efficiency* of the antenna, or the ability of the dish to capture the received signal. A perfectly designed dish will have 100 percent efficiency in that it will capture 100 percent of the received signal. As a result, a dish antenna with a high efficiency will also have higher gain than a dish with low efficiency. Factors that affect the efficiency of a dish antenna include the accuracy of the reflective surface, blockage of the reflective surface by the feed, signal loss caused by the absorption of some microwave energy into the dish antenna surface; and signal losses that occur as the signal travels between the feed and the amplifier.

We can refer to those factors as surface accuracy, profile accuracy, and physical size. Surface accuracy provides a measure of the RF smoothness of the reflecting surface of the antenna. If the antenna has a protective coating the reflecting surface will not show. Profile accuracy represents a measure of how closely the shape of the dish approximates a true parabolic shape. The importance of RF smoothness and profile accuracy increases with frequency. As the frequency of the received signal increases, the effect of small surface deviations becomes more significant with respect to the wavelength of the signal.

Physical size remains important for two reasons. While larger dishes provide greater gain, the reflectors arrive in sections, or petals. Surface and profile accuracy depends on the supporting struts for those petals. As a result, the possible deformation of the dish profile can become more likely with larger dishes. In satellite antenna design, a trade-off occurs between the gain and the efficiency of the antenna. If we change the efficiency of the reflector from 65 percent to 50 percent, the gain will decrease by 1.1 decibels. Increasing the dish size stands as the only method for compensating for this loss in a large reflector.

As the *Table 3.3* shows, antenna efficiency increases slightly and then decreases as the size of the reflector antenna increases.

Diameter	Gain	Efficiency	Feed and Dish Construction
1.6 meters	34	56%	Prime Focus, Solid
1.8 meters	35	56%	Prime Focus, Solid
3.0 meters	40.2	60%	Prime Focus, Mesh Petals
3.65 meters	42.3	66%	Prime Focus, Mesh Petals
5.0 meters	44.5	58%	Prime Focus, Mesh Petals

Table 3.3. Comparison of reflector gain and efficiency by size C-band.

Main and Side Lobes of a Dish Antenna

We can calculate beamwidth as a ratio of the main lobe of the dish to its side lobes. In terms of physical measurements, beamwidth is the width of the main lobe between points on the reflective surface where power has dropped by 3 decibels. *Figure 3.2* illustrates the difference between the main lobe and side lobes. Referring to the figure, the main lobe projects out towards the region that has the maximum amount of power. Side lobes project off to either side of the main lobe and indicate the capability of the dish antenna to pick up microwave signals from off-axis sources.

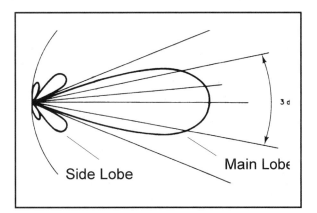

Figure 3.2. Main and side lobes of a dish antenna.

If the dish antenna has the proper design, the main lobe will appear much larger than the side lobes. Yet, the width of the main lobe must remain consistent with the requirement to have a narrow bandwidth. Since commercial satellites have a 2° spacing between orbital slots, the magnitude of the side lobes becomes increasingly important. Side lobes located within this 2° range and with enough power would allow the dish antenna to receive signals from the adjacent as well as desired satellites simultaneously.

Dish Antenna Beamwidth

If we view the dish antenna in terms of a telescope pointed at a particular section of the sky, it becomes easier to say that the dish has an established field of vision. When we viewed the antenna gain equation, we calculated the gain for the main axis of the main lobe of the antenna. As shown in *Figure 3.3*, the gain of the antenna drops away from the main axis.

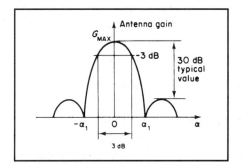

Figure 3.3. Illustration of gain at the main axis of a reflector antenna.

Beamwidth defines the preciseness of the region that the dish antenna can target. A dish antenna that has a narrow beamwidth becomes less susceptible to interference from adjacent satellites. A wide beamwidth allows the simultaneous reception of signals from the desired satellite and signals from the adjacent satellites. The signals from the adjacent satellites would appear as interference.

Referring back to *Figure 3.2* and also to *Figure 3.3*, the equation for finding the half-power beamwidth of an antenna appears as:

$$q \text{ degrees} = k \, l \, / \, D$$

where:

q degrees = 3 dB beamwidth (degrees)
l = wavelength (meters)
D = earth station diameter (meters)
k = the aperture illumination factor

When an antenna has a uniformly illuminated aperature, the aperature illumination factor has a range between 58 and 65.   If the antenna has a tapered aperature, the illumination factor will have a value of 70.

Dish Antenna Materials

As mentioned, the materials used to manufacture a dish antenna vary from aluminum and steel to fiberglass and wire mesh.   While steel, fiberglass, and wire mesh dishes gained usage during the 1980's, most manufacturers have opted for solid or perforated aluminum dishes.   The lightweight aluminum dishes allow easier construction and provide lower shipping costs.   Older steel dishes gave superior durability in terms of wind and hail damage but did not gain popularity because of weight, cost, and the danger of oxidation.   Wire mesh dishes yielded lower costs at the expense of strength and performance.   Well-constructed fiberglass dishes include an embedded wire mesh and provide superior performance when compared to aluminum dishes.   However, the popularity of fiberglass dishes has decreased because of manufacturing costs and appearance.

Solid vs. Mesh Construction

Any satellite dish antenna must contain some type of metal reflector.   Without the metal portion of the dish, microwave signals would pass through the material.   As a result, the dish antenna could not capture and reflect the signals towards the amplifier circuitry.   Given the wavelength of C- and Ku-band signals, the metal reflector must have either a solid composition or feature perforations that have a smaller diameter than the signal wavelength.

Non-solid dishes either have a perforated aluminum construction or a wire mesh construction and are used mostly in C-band installations.   While a perforated dish has fewer, larger holes, a mesh dish has many fine holes.   Although perforated and mesh dishes do not have the gain seen with solid dishes, the loss in gain is minimal. More than anything, mesh or perforated dishes present a cleaner aesthetic appearance than the solid dishes. *Figure 3.4* shows a reflector antenna constructed from wire mesh materials.

Figure 3.4. A reflector antenna
constructed from wire mesh.

Dish Antenna Manufacturing Processes

Three processes exist for the manufacture of reflector antennas. The first involves placing a flat sheet of metal into a parabola-shaped mold and then spinning the entire assembly. As the assembly spins, a shaping roller forms the sheet metal into a parabola. In contrast to the spun process, another manufacturing technique forms a parabolic dish by using a hydraulic press and the combination of a male and female die to press the sheet metal into a parabola.

Fiberglass and PVC dishes form through an injection molding process. Because fiberglass or plastic alone cannot reflect microwave signals, the process begins with the placement of metal foil into a parabolic-shaped mold. Injection molding of the heated, fluid-like fiberglass or plastic encapsulates the metal foil within a parabolic shape and forms the reflector antenna.

Reflector Types

Reflector antennas vary by type as well as size or construction method. While most consumer installations rely on parabolic reflectors and the prime focus method, some direct satellite systems utilize a parabolic reflector with an offset-focus. Commercial installations may use a parabolic reflector with a prime focus, a parabolic reflector with an offset-focus, a Cassegrain reflector, or a spherical antenna.

Parabolic Dish Antennas

Considered as a standard in the satellite industry, the parabolic antenna takes its name from the curve that causes the entire received signal to bounce into a central, well-defined focal point found above the antenna. The accuracy of the parabolic curve allows the dish to deliver a greater amount of signal to the feed assembly. Generally, the outer perimeter of a parabolic antenna has a circular shape. To achieve the optimum reception, a parabolic antenna must point directly at the desired satellite. *Figure 3.5* shows a parabolic dish.

Figure 3.5. A solid parabolic dish.

Prime Focus Dish Antennas

Most parabolic antennas utilize the prime focus method for mounting the feedhorn/amplifier assembly on the dish. Illustrated in *Figure 3.6*, the prime focus method either uses a tripod assembly for the antenna feed or a buttonhook assembly. As the name "prime focus" indicates, either method positions the assembly over the center of the dish at the distance required for maximum concentration of the signal. Consequently, the design reflects any signals received from locations other than the main axis of the dish away.

Figure 3.6. A prime focus feed assembly.

Offset-Fed Dish Antennas

Although parabolic dish antennas have gained widespread usage, a few negative characteristics exist with the design. With the feed mounted at the focal point, it may also pick up stray noise generated when the dish reflects signals from other points than the main axis. Because microwave signals are examples of electromagnetic waves, the signals received from the satellite not only strike the main axis but also spread around the dish and the feed. If we could see the microwave signals, we would see diffraction of the signals as well as reflection.

A dish antenna constructed as offset-fed antenna uses only a section of the parabola seen with the prime focus dish antenna. As a result, the feed seems to set at an offset angle from the dish antenna. The shape of the dish antenna and the placement of the feed establish a design where the feed does not block any portion of the antenna surface. With the feed facing more into deep space and less into the ground, the noise temperature of the dish decreases. The construction of the offset-fed dish also places the dish antenna at a steeper angle with respect to the satellites in the Clarke Belt. Moving to *Figure 3.7*, an offset-fed antenna also features a shorter focal length and smaller dimensions than those seen with the parabolic dish antenna.

Figure 3.7. Offset-fed dish antenna.

Offset-fed dish antennas have greater efficiency since the feed, amplifier, and feed supports do not block the signal as it strikes the reflective surface. Moreover, an offset-fed dish antenna will have smaller side lobes than other types of dish antennas. The offset-fed design also has lower noise temperatures because the feed points away from the ground at a steeper angle.

Cassegrain Dish Antennas

Shown in *Figure 3.8*, the Cassegrain feed antenna design mounts a subreflector at the focal point of the dish and the amplifier assembly at the center of the dish. As a result, the subreflector sends a more concentrated signal to the amplifier. Typically used with larger dish sizes, the Cassegrain feed offers a slight gain in antenna efficiency. With the amplifier placed in the center of the dish, Cassegrain antenna designs must feature some method for preventing moisture from entering the feed. Because of an increased susceptibility to adjacent satellite interference, Cassegrain antennas must have precise alignment.

Figure 3.8. Cassegrain feed antenna.

Spherical Dish Antennas

Depicted in *Figure 3.9*, the curvature of a spherical antenna creates a sphere rather than a parabola. Because of this shape, a spherical antenna has a separate focal point for each satellite, opposed to a parabolic antenna with a single, well-defined focal point. This provides an advantage for some commercial installations in that the spherical antenna can "see" a large portion of the Clarke Belt while remaining fixed in one location.

Figure 3.9. Depiction of the curvature of a spherical antenna.

Determining the Best Location

The microwave signals that carry the information from the satellite to the dish reflector exhibit the same characteristics as light with the exception of visibility. As such, those signals travel in a straight path along the line of sight. With the geosynchronous satellites positioned over the equator and the satellite system in the northern hemisphere, the dish must have an unobstructed view of the southern sky. Any tall buildings, power poles, trees, or other solid objects will prevent the signals from reaching the dish.

A simple site survey may consist simply of the consumer or installer standing near the point of the desired location and looking towards the south. When installing a C-band system driven by an actuator, a clear and unobstructed view of the sky from the southeast to the southwest. While conducting the site survey, take into consideration the season of the year. During the late fall, trees without their foliage may not seem to obstruct the signal.

Using an EIRP Map

Downlink antennas on the satellite can transmit signals to any geographic region within direct view of the satellite. The transmission of signals from the satellite forms a particular shape called the *footprint* at the geographic area covered by the transmission. Referring to *Figure 3.10*, the signal remains strongest at the center and diminishes towards the outer edges. Every satellite has a characteristic footprint that depends on the amplification of the signal by the transponders, the satellite antenna, the satellite orbit, and the inclination of the satellite.

The footprint map provides the information that we need for selecting the proper size for a reflector antenna. Looking at the map, each circle represents a different level of effective isotropic radiated power, or EIRP. As with other units of power, we express the EIRP in decibels relative to one watt. With EIRP, we have equal distribution of power in all directions if the antenna equally radiates in all directions. A transponder with limited power can have a high EIRP in one direction if the antenna radiates at a higher level in that direction.

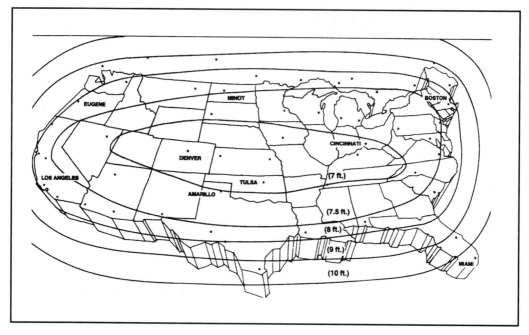

Figure 3.10. An EIRP map.

The circles on the EIRP map represent areas that have equal levels of EIRP. Smaller circles have higher EIRP levels but smaller geographic areas. Larger circles have lower EIRP levels but larger geographic areas. As we move from circle to circle, the power level changes for several reasons. Rather than measure EIRP levels at the earth's surface, we measure the levels at the satellite downlink antenna. With the antenna concentrating the signal towards a given location, the signal spreads out as it travels to earth and experiences *free space loss* because of the distance covered by the signal. In addition to the degradation caused by distance, the signal also loses gain as it travels through the earth's atmosphere. Water vapor in the atmosphere blocks the signal.

In mathematical terms, free space loss, or L_{FS}, appears as a function of the following equation:

$$L_{FS} = 20 \log [(4000pD)/l] \text{ (dB)}$$

where:

p = 3.14159,
D = 23,000 miles or the distance between the satellite and the earth, and
l = wavelength in meters

Using a Signal Strength Meter

Although we can spend time using the dish to hunt for satellite signals while looking through the living room window at a television, the best method for accurately aligning the dish is the use of a signal strength meter. Shown in *Figure 3.11*, the signal strength meter monitors the first intermediate frequency signal that occurs after downconversion by attaching in series with the LNB and the transmission cable. Low-cost signal strength meters indicate peak signals through a simple meter or audio tone.

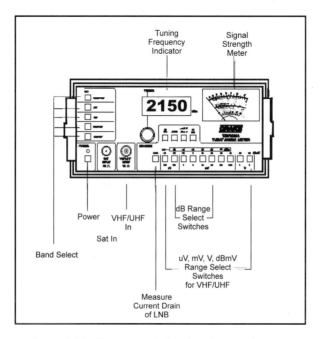

Figure 3.11. Front panel of a signal strength meter.
Courtesy of Drake Electronics.

More expensive signal strength meters supply the LNB with internal batteries and have more sophisticated digital displays. The external supply for the LNB provides protection for the receiver while the system is under test. User-switchable attenuators found within the test equipment allow the setting of the sensitivity and provide a relative indication of signal strength. The more advanced signal strength meters also allow the measurement of the C/N ratio for each channel, have a direct frequency input and a composite video output, and have microcomputer control. Given those additional features, the test equipment provide for the selective identification of desired satellites and the peaking of polarization.

Checking for Sources of Terrestrial Interference

As we saw in chapter one, microwave interference from telephone transmission stations can interfere with the reception of satellite signals. Good practice involves testing for and blocking any type of terrestrial interference before beginning the installation. In many instances, we can shield the system from microwave interference by installing the dish behind trees or buildings. Depending on the location, we could also install the dish closer to the ground or within a small depression. In extreme cases, we can use wire mesh as a shielding material or we can install notch filters that suppress the interference.

In many instances, we can avoid physical obstructions and gain good line-of-sight by placing the reflector on a higher pole. However, the risk for increased terrestrial interference also increases with the height of the reflector antenna. Before making the final decision about a pole mount, test for microwave interference at the same height as the dish.

Three basic methods exist for checking for the presence of terrestrial interference. Many installers will use a portable satellite system to test a proposed site. With this method, the installer will place the test system as close to the proposed site as possible. In addition, this method requires that the installer direct the dish to at least four or five satellites along the Clarke belt while testing the reception on all transponders.

The second method is a variation of the first. Rather than employing an entire system to check for terrestrial interference, we can connect the combination of an LNB and feedhorn to a satellite receiver, signal strength meter, and television. After initializing all the electronics, we can then scan the LNB/feedhorn combination in all directions. The television screen should show white noise, or snow, and the signal strength meter should remain unchanged if no terrestrial interference is present.

If the television raster changes from white noise to either black, blank, white bars, black bars, or white lines and the signal strength meter exhibits a positive reading, terrestrial interference exists. Moving the LNB/feedhorn combination back and forth and watching the television raster and signal strength meter will disclose the direction of origin for the interference. *Figure 3.12* depicts the use of the LNB/feedhorn combination used to test for terrestrial interference.

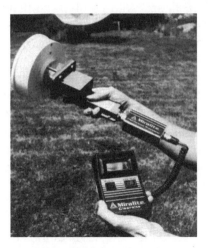

Figure 3.12. LNB-feedhorn
terrestrial interference test
combination.

Spectrum Analyzers

The third method utilizes a more sophisticated piece of test equipment called a spectrum analyzer. With a spectrum analyzer, we can see the signals attempting to enter the satellite system. Spectrum analyzers display levels of detected signals found across the satellite frequency band. This type of equipment shows the frequency of the terrestrial interference as well as the power of the microwave energy coming from any direction. Although expensive, spectrum analyzers offer adjustable frequency ranges and high signal sensitivities.

In the not too distant past, only engineers and senior electronic technicians utilized spectrum analyzers. Now, most satellite system installers have access to a spectrum analyzer. The device contains special filters along with linear and logarithmic detectors that perform signal processing. In the linear mode, the device references all voltage readings from the bottom of the screen at zero volts at a volts/div sensitivity. In the more commonly used logarithmic mode, the device references occur at the top of the screen in decibels. We can calibrate the readings in dbm, or decibels to one milliwatt, or in decibels—the ratio of one signal to another signal in volts or power.

During operation, the output of the spectrum analyzer appears on an oscilloscope CRT. Part of the decision process in operating a spectrum analyzer lies within selecting the best bandpass filter for the best job. When using the spectrum analyzer, the operator selects a bandpass filter after the IF stage that has a bandwidth narrower than the bandwidth of the IF filter. With its characteristics carefully chosen by the manufacturer, the bandpass filter acts as a resolver of frequencies within the translated frequency range of the intermediate frequency.

Any signal that has the same frequency as the bandpass filter passes onto the next signal processing stage within the spectrum analyzer. In addition, the analyzer also passes along information about the strength and amplitude of the signals. Other filter circuits translate and identify the signals coming from the antenna. With this information, the operator can calibrate the analyzer in both frequency and amplitude and have an analysis of the frequency and strength of the incoming signal.

All this seems complicated. However, the operator does not need to know the actual frequency of operation. Instead, the operator concentrates on the bandwidth measured at specific levels relative to the peak level of the center frequency of the bandpass filter. A filter with a 10 KHz bandwith has the capability for resolving signals 10 KHz apart from each other. As shown in *Figure 3.13*, the oscilloscope display will show a 3 dB notch between the two signals. With narrower bandwidths, we can check a signal for power line hum at 60 Hz or 120 Hz.

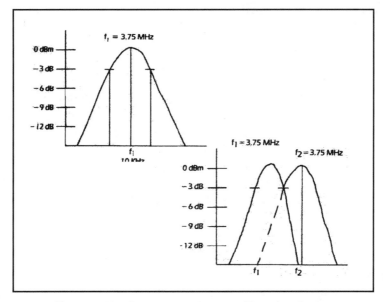

Figure 3.13. Spectrum analyzer oscilloscope display.

However, as we choose bandpass filters with narrower bandwidths to resolve various frequencies, the spectrum analyzer takes longer to provide a measurement. In effect, the device electronically sweeps across the selected band while attempting to detect all signals within its view. Choosing a wide band of frequencies increases the sweep rate of the signal and clutters the display. Most spectrum analyzers automatically select the best sweep rate and apply the best filter for the task.

The key to using a spectrum analyzer rests within the operator's ability to interpret the display. When we measure the amplitude of signals, each of those signals have different amplitudes with respect to one another. Depending on the signal type, the difference in amplitudes may form a ratio of 10,000,000 to one. An oscilloscope trace may display low level signals or high level signals but not a combination of both. The strong signal will appear at the top of the display with the weak signal off the display. Logarithmic amplifiers or log amps attached to the spectrum analyzer compensate for the amplitude differences and allow the simultaneous viewing of both signals. A spectrum analyzer used for measuring TVRO signals should have a 40 dB (10,000:1) to 60 dB (1,000,000:1) dynamic range capability.

Eliminating Terrestrial Interference

If we discover terrestrial interference at an installation site, we have four options:

•	Select an alternative site;

•	Build screens to block the TI;

•	Add a waveguide filter to trap the interference at the 3.7 to 4.2 GHz frequency

•	Add filters on the 70 MHz line between the LNB and the receiver.

As mentioned before, we can eliminate terrestrial interference with screening. The number and type of shields depends on the severity of the terrestrial interference and the number of directions from which the terrestrial interference arrives. If the problem requires elaborate screening, we can camouflage the screening so that it resembles a portion of a fence, gazebo, or trellis.

Selecting the Best Reflector Size

During an earlier section of this chapter, we found that EIRP footprint maps show the power levels of signals for specific geographic areas. The data obtained from those maps assists with the size selection of the reflector antenna. Because of breakthroughs in LNB and receiver technologies, an eight-foot reflector will serve well in almost any location for C-band reception. Lower temperature LNBs and improved amplifiers within the receiver compensate for the loss of gain given through the use of a smaller diameter dish. When viewing an EIRP map, the required size of the reflector antenna decreases with increases in EIRP level. If the carrier-to-noise ratio for the particular site increases, the size of the reflector antenna also increases.

We can use the combination of EIRP levels to find the optimum combination of antenna size and LNB noise temperature needed to receive a signal that exceeds the minimum threshold requirements of the receiver. *Figure 3.14* provides a cross-reference for C-band EIRP levels, LNB noise temperature, and antenna reflector size. Referring to the table, the intersection of the bold EIRP line with dashed antenna size line above the dotted LNB line discloses that resulting signal level will exceed a signal level of 7 decibels. Sparkle-free reception of signals would require a two-decibel margin above the receiver threshold.

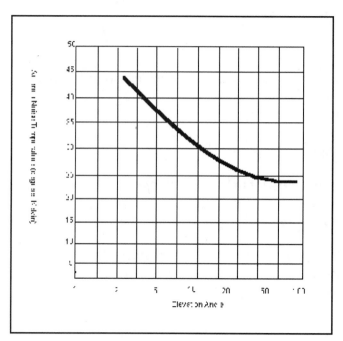

Figure 3.14. Cross reference between EIRP levels, noise temperature, and reflector size.

Summary

Chapter three provides a detailed overview of dish antennas and the theory behind dish antenna operation. Within the chapter, you found theoretical and practical information about dish antenna design and construction. The information introduced and defined concepts such as azimuth, elevation, and focal length. In addition, the chapter provided information about selecting the correct size for a reflector and the best location for an installation. The information also covered the use of signal meters and spectrum analyzers.

The chapter continued with this emphasis on location and installation with a discussion about terrestrial interference and methods for blocking TI. As you move into chapter four, the discussion about terrestrial interference will become valuable as you work through the practical steps for installing a dish antenna. Chapter three concludes with an overview of EIRP maps.

CHAPTER

4

Moving from Definition to Assembly

Chapter four begins to introduce the individual parts that make up a satellite reception system. The chapter opens with a detail description of the dish antenna and the factors that affect the performance of the antenna. From there, the chapter moves to a discussion of the materials typically used to construct dish antennas and an overview of dish antenna configurations. The chapter covers how to find the proper location for the dish, assemble the mount and ground pole, align the dish antenna and antenna mount, and assemble the feedhorn. In addition, the chapter includes an overview of different types of antenna and dish installations.

Installing C-band system hardware involves the cementing of a pole into a hole dug at the installation site, mounting a support pole to a concrete slab, or installing a dish support to a roof of a building. Regardless of the type of installation, the pole must have an absolutely vertical orientation. In addition, the installation also calls for trenching for conduit that will protect the transmission cables, assembly of the dish mount, and assembly of the dish. Each portion of the installation process requires careful planning and consideration of the need to properly align the dish antenna with satellites orbiting in the geosynchronous arc.

After completing the site survey, planning for the installation should cover not only the main assembly steps for the dish, mount, and feedhorn and signal strength requirements but also the need for proper tools and a workspace. Moreover, careful planning should also consider the location of the paths for the transmissions, the protection of the dish hardware and electronics from the elements, and aesthetics. When preparing to dig trenches for the transmission cables, always ensure that the trench avoids water and electrical lines. The location of the dish should allow some protection against wind loading to occur and should include proper grounding in the event of a lightning strike. Aesthetic appeal considers both the opinions of the home owner and any local ordinances.

Assembling the Reflector Antenna

Most manufacturers enclose a set of assembly instructions with the reflector antenna. The instructions should list any required tools, list all parts, and illustrate the assembly of the dish. As you may recall, we can purchase different types of reflector antennas that include spun aluminum or steel reflectors, fiberglass reflectors, mesh reflectors, and perforated reflectors. While the spun aluminum or steel reflector antennas arrive in one piece, the other categories require the assembly of individual petals and a cover point for the center of the dish. Most have pre-drilled holes for the mount.

One-Piece Aluminum or Steel Reflectors

With the reflector arriving in one piece, assembly becomes nothing more than bolting the mount to the antenna. However, we should remember to carefully tighten the bolts while fastening the two together. Applying too much torque to the mount bolts could cause the dish to have a deformity around the mounting hole. In addition to carefully tightening the bolts, we should also ensure that the mounting holes are centered. To check the centering of the mounting holes, tie strings at opposite corners of the mount and then intersect the strings across the dish. The intersecting point of the strings should align with the center of the dish.

As with the mount bolts, also exercise care when fastening the center plate to the dish antenna. Any overtightening of the bolts could deform the surface of the dish. With either the mount bolt or the center plate bolt locations, deformities at the dish surface will cause a signal loss. In addition, always use all washers, lockwashers, or rubber spacers as provided and intended by the manufacturer. The rubber washers prevent a chemical reaction called electrolysis from occurring between two different types of metal.

Fiberglass, Solid Metal, Mesh, and Perforated Reflectors

Fiberglass and solid metal reflectors usually arrive in sections or petals. Because of the need for a smooth, well-formed parabolic reflective surface, the assembly procedure requires that the rim of the dish remain flat. Many times, the use of a cardboard box or plastic wastebasket as a support while assembling the petals can save time and protect the surface of the dish. In addition, always assemble the reflector on a padded surface such as a carpet, grass, or cardboard sheet.

When assembling the petals, do not tighten any bolts until after completing the entire assembly. As you tighten the bolts from petal to petal, always check to ensure that the rim of each section remains flat and that surface of each petal remains smooth in reference to the other petals. *Figure 4.1* shows the assembly of the petals.

Figure 4.1. Assembling the petals of a solid reflector antenna. Courtesy of Channelmaster, Inc.

As with the fiberglass and solid metal reflectors, mesh and perforated reflectors arrive in separate petals. Mesh reflectors require even more special handling because of the possibility of denting the mesh surface or of having the entire panel popping out from its frame. Any installation that utilizes a mesh reflector will require some maintenance of the petals. Perforated reflectors have a stronger construction than mesh reflectors.

Regardless of the type of dish antenna used, always complete the reflector assembly by checking the dish for warpage. Again using the string method, fasten four strings at opposite corners of the antenna and then stretch the strings across the dish. The strings should intersect only at the center of the dish and should barely touch one another. If the strings intersect at different points, have a lot of separation, or come together tightly, warpage of the reflector surface has occurred. *Figure 4.2* illustrates the stringing of the dish antenna. Generally, we can loosen and retighten bolts at either the petals or the mounts to lessen the warpage.

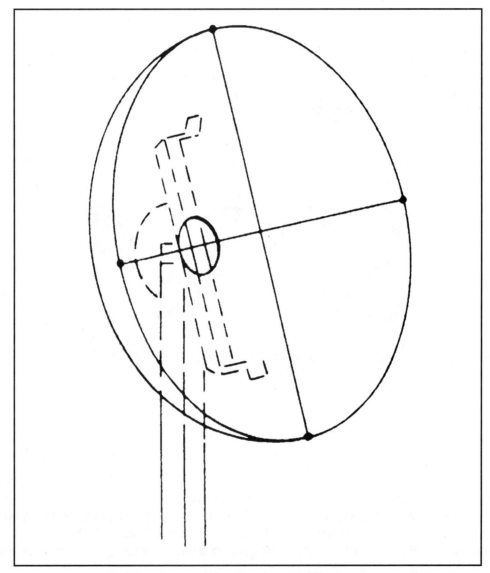

Figure 4.2. Illustration of stringing the dish antenna.

Dish Antenna Mounts

Most consumers overlook the importance of the mount that supports the antenna. The steel mount carries the full weight of the dish and maintain the precise position of the dish with regard to a specific satellite. While this may seem simple, maintaining the precise position must occur in all weather conditions. Any change in alignment— even a change as small as a fraction of an inch—can make the difference between the reception of high quality audio and video and no reception.

Aligning the Dish and Mount

Proper alignment of the dish and mount stands as one of the key factors that will affect the overall performance of the satellite receiving system. Alignment of the dish and mount involves finding the polar axis and setting the elevation of the dish. Without the proper installation and alignment of the dish and mount, even the best receivers or low-noise amplifiers will give poor performance.

Setting the Azimuth and Elevation for Your Location

The *azimuth coordinate* of the satellite system represents the bearing of the satellite from the location of the installation. In turn, the *elevation* or *polar angle* represents the angle at which the dish looks up at the satellite. When working with a C-band system that tracks a number of geosynchronous satellites, every satellite within view of the installation has unique azimuth and elevation coordinates. The latitude and longitude of the system location affect the azimuth and elevation for each satellite tracked along the arc.

Finding the Polar Axis

The *polar axis* exists in parallel with a line passing through the north and south geographic poles and varies with the location of the satellite receiving station. If we have a location where the latitude of the installation equals 23° North, the polar axis equals 23° as measured from the horizon.   In any case, the polar axis must orient exactly along a north/south line.

By knowing the azimuth coordinate of a specific satellite, we can use a compass to point the dish antenna in the correct direction. When using the compass, step away from any large metal objects or ac power lines and transformers. First, we must correct the compass readings for true north. If true north rests east of magnetic north, we subtract the correction factor from the compass readings. If true north rests west of magnetic north, we add the correction factor to the compass readings.

If the satellite receiving system is located in the Northern Hemisphere, the satellite orbiting closest to true south will have the highest point or elevation of all the satellites. For systems located in the Southern Hemisphere, the satellite found closest to true north will have the highest point or elevation of all the satellites. Satellites that have a bearing farther away from due south will have lower elevations. At some point, the elevation for satellites to the extreme west or extreme east of the system location will fall below the horizon and will prevent reception.

Using an Inclinometer

To accomplish this, use an *inclinometer*, or angle finder, to measure the elevation angle of any dish antenna and to ensure that no obstructions could block satellite signals during a site survey. We can position the inclinometer on a level surface at the center of the site and line it up with the bearing of a satellite. Then, we simply tilt the inclinometer back until it indicates the elevation of the desired satellite.

Finding the Polar Angle

Adjusting the polar angle of the antenna mount involves both simplicity and precision. The *polar angle* of the mount equals the latitude for your location. We can measure this angle by using an inclinometer placed on the back of the mount as shown in *Figure 4.3*. The offset angle adjustment on the mount allows the setting of the polar angle. Establishing the polar angle along with orienting the mount to a position parallel with the north/south axis of the earth allows the antenna to symmetrically track the Clarke Belt.

The Declination Offset Angle

The *declination angle* measures the amount of tilt of a polar mount away from the Earth's axis of rotation. When installing a satellite receiving system, adjusting the *declination offset angle* decreases the sighting angle of the reflector antenna so that it focuses on the satellites in the Clarke belt and must occur within 1/2-degree accuracy. Too much or too little declination-offset angle will cause the antenna to miss the Clarke Belt. As shown in *Chart 4.1*, declination offset angles increase as the location for the system moves farther away from the equator. *Figure 4.4A* illustrates the difference between applying only the polar angle and the application of the polar angle plus the declination offset angle.

Figure 4.3. Using an inclinometer to set the polar angle of a dish antenna.

Latitude (degrees)	Offset Angle (degrees)	Conversion Factor	Polar Axis (degrees)	Elevation Angle (degrees)
5	0.76	.0157	5.13	5.89
10	1.51	.0310	10.26	11.77
15	2.25	.0454	15.37	17.62
20	2.98	.0611	20.47	23.45
25	3.66	.0751	25.57	29.23
30	4.33	.0875	30.63	34.96
35	4.97	.0998	35.68	40.65
40	5.56	.1104	40.71	46.27
45	6.11	.1210	45.71	51.82
50	6.62	.1298	50.69	57.31
55	7.06	.1378	55.66	62.72
60	7.47	.1441	60.59	68.06
65	7.80	.1495	65.52	73.32
70	8.09	.1512	70.43	78.52
75	8.31	.1548	75.33	83.64
80	8.47	.1548	80.22	88.69

Chart 4.1.

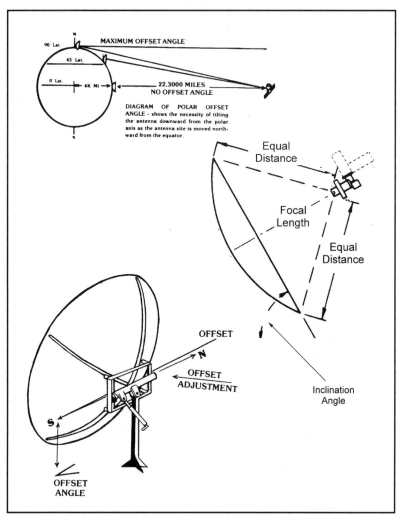

Figure 4.4A. Illustration of offset and inclination angles.

Mathematically, the declination offset angle equals:

$$90 - \text{arc cosine} \ (R \ (\text{Sin} \ Y) \ / \ Z \ )$$

where:

R = 4,000 miles (radius of the earth)
Z = Satellite range (distance) in miles or 26,300 miles
Y = Site latitude in degrees

However, easier methods exist for finding the declination angle. As the following shows, we can rely on declination offset angle charts, inclinometers, and even tape measures when setting the offset angle.

As with the polar angle, we use an inclinometer to measure the declination offset angle. To set the angle we find the declination angle listed in the chart and then add that number to the site latitude. With the system located at the equator, we do not use an offset angle. However, as we move farther south or north of the equator, the installation requires more tilt. *Table 4.1* also provides both the angle and a conversion factor that coverts the angle to inches. Through the use of the measurements given in the chart, we could set the mount with a tape measure.

To set the offset angle with an inclinometer we place the device on the rear flat surface of the dish antenna that extends in parallel to the line between the rims. With everything in place, we then adjust the declination angle bolt on the mount until the inclinometer shows the latitude plus the declination angle. The declination bolt usually consists of a threaded bolt that fastens to the reflector and extends through a plate on the mount. As shown in *Figure 4.3*, nuts on either side of the declination plate allow the adjustment of the angle.

To use only a tape measure, we must rely on the factors shown in the chart. For example, an antenna site located at 35° latitude will have a factor of 0.0998. With this factor in mind, we measure the distance between the upper and lower pivots on the mount and multiply the two measurements. If the distance between the two points equals 28 inches, the multiplication of 28 times 0.0098 should equal 2.79 inches. Knowing this, we can adjust the offset so that it equals 2.79 inches plus the length of the lower bracket.

Regardless of the method used to set the offset angle, the angle varies approximately 1/2 degree as we swing the reflector and mount through an arc ranging from 79° to 143° West. With most antennas having a two-degree beamwidth, the 1/2-degree variation does not cause any problems. As *Figure 4.4B* shows, however, any misalignment of the reflector and mount by 1/2 degree or more can cause a significant accumulative error that will readily affect the performance of the system.

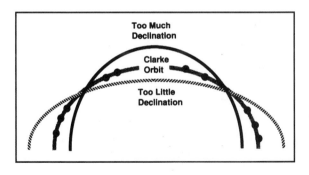

Figure 4.4B.
Symptoms of reflector
and mount
misalignment.

Azimuth-Elevation Mounts

An azimuth-elevation, or AZ-EL, mount matches the azimuth and elevation of the satellite through a series of adjustments but does not allow the re-orientation of the dish to another satellite. Moving the dish to a new satellite requires the readjustment of both the azimuth and the elevation of the dish.

Referring to *Figure 4.5*, the azimuth adjustment occurs through the use of a small motor and moves the dish mount to the correct azimuth angle for the desired satellite. Then, another motor moves the dish mount to the correct elevation for the reception of the satellite signals. The value of AZ-EL mounts becomes apparent through the use of the mounts in mobile satellite reception systems. Regardless of the location or orientation of the vehicle or vessel carrying the mount, the motor adjustments allow the precise targeting of the desired communications satellite.

Figure 4.5. An AZ-EL mount.

Installing an AZ-EL Mount

Given the simplicity of the AZ-EL mount, installation involves only a few steps. Before beginning the installation, obtain the necessary azimuth and elevation angle measurements for our location. Always remember to correct the azimuth angle for magnetic deviation. As shown in *Figure 4.5*, the mount attaches to the reflector antenna through four bolts. The mount sleeve fits over the ground pole and fastens firmly through the use of several mounting bolts. Tighten the bolts to the point where the reflector and mount only turn if pressure is applied. AZ-EL mounts clamp to a specific azimuth and elevation setting which allows the reflector to focus only on one satellite. Use either an inclinometer or stamped graduations found on the mounting bracket to set the proper elevation angle.

Connect a short coaxial cable between the LNB and a signal strength meter. Then, swing the antenna and mount to the correct azimuth angle. As you approach the correct azimuth angle, the signal detected by the meter should increase. Move the antenna and mount slightly until the signal reaches its maximum level. Then, fine tune both the elevation angle and the azimuth angle to achieve peak signal performance.

Polar Tracking Mounts

A polar mount allows a dish antenna to rotate in an arc and track satellites orbiting in the Clarke Belt. When installing a polar mount, the axis must align with true south for installations located north of the equator and true north for installations located south of the equator. Once the dish antenna has the proper elevation, it will track all geostationary satellites in view of dish through an adjustment in direction. As a result, a dish antenna attached to a polar mount automatically changes elevation as the dish moves to the east or west. *Figure 4.6* shows a typical polar mount assembly while *Figure 4.7* depicts the relationship between the polar mount and the Clarke Belt.

Figure 4.6. A polar tracking mount.

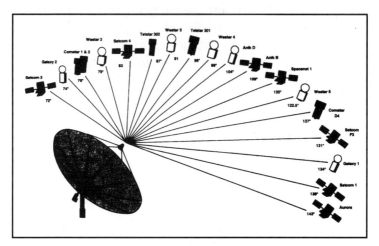

Figure 4.7. Relationship of the polar mount and the Clarke Belt.

Installing a Polar Tracking Mount

As we saw in chapter two, a polar tracking mount targets a desired satellite by rotating around an inclined axis. Because the polar axis of the mount tilts away from parallel to the Earth's axis of rotation, the movement around the axis causes the dish antenna to sweep along an elliptical curve that crosses the equatorial plane that matches the circular belt created by the geosynchronous satellites. To properly target a satellite, the pivotal mount of the polar mount must accurately align with true north. Slight inaccuracies in this alignment may result in uneven tracking along the geosynchronous band.

Installing a Horizon-to-Horizon Mount

Horizon-to-horizon mounts feature installation methods seen with both the AZ-EL mounts and the polar mounts. Moving to *Figure 4.8*, the mount attaches to the reflector antenna through four bolts located near the center of the dish. To set the polar elevation for the mount, we adjust the elevation angle plate located near the ground pole sleeve. An upper declination bracket located near the back of the reflector allows the fine adjustment of the declination angle. The horizon-to-horizon mount operates as a direct-drive unit from a motor and a gearbox. With this arrangement, the mount can track a full 180-degree azimuth.

Installing the Reflector and Mount at the Pole

The next task in the assembly process involves fastening the mount to the pole and the reflector antenna to the mount. While fastening the mount seems basic, we need to remain aware of the alignment needed for the mount to properly focus the reflector on desired satellites. Once we have aligned the reflector and mount, we also need to ensure that the bolts fastening the mount to the pole remain firmly tightened. A good practice for keeping the mount aligned after tightening the bolts involves the drilling of a small hole through the mount sleeve and the pole and the insertion of a lock pin into the hole.

Figure 4.8. Horizon-to-horizon mount.

When preparing to fasten the reflector to the mount, consider the weight and bulk of the reflector antenna. More than likely, two individuals can easily handle any reflector antennas that have a diameter of less than 12 feet. Installation of the dish to the mount begins so that the positioning of the mount accommodates the easy attachment of the mount bolts. Two choices for positioning exist. With the first, we can position the mount so that it remains parallel with the pole. The positioning of the mount in this way allows us to roll the dish into place. Our second option involves placing the mount perpendicular to the pole as shown. With this, we can lift the dish onto the mount, place it in a position similar to a birdbath, and then attach the mounting bolts.

Regardless of the method used to install the dish onto the mount, we should ensure that the mount cannot pivot. Unless fastened securely, the dish and mount can swing freely from side-to-side. Taking precautions such as this will allow the installation to occur without mishap and prevent damage to the reflector or injury to the installers.

Setting the Dish Support Pole

Several different options exist for setting the dish support pole. One of the most common involves the setting of the pole in a ground mount where the installer digs a hole for the pole support and then sets the pole with concrete. Another common method involves either the running of a concrete slab or concrete pads for a three-legged support platform. Along with those methods, we can also mount a C-band dish support on a roof . To accomplish the latter, we can use a peak roof mount, a pitched roof mount, or a flat roof mount. When considering any of the roof mount options, always remember that the installation of the dish antenna will result in both wind and snow loading. Regardless of the support option used, always remember to allow for the cabling runs that accompany in system installation.

Setting a Ground Pole

For most C-band installations, setting a ground pole provides the best method for supporting the dish antenna. With this method, we dig a hole that has a diameter at least two to four times the diameter of the pipe used to support the dish and at least three feet into the ground. Generally, a ground pole will extend five feet above ground. If the pole extends more than five feet above ground for a particular installation, deepen the hole by at least six inches for every 24 inches in added height.

Ground Pole Wind Loading and Recommended Diameters

Wind loading stands as another important issue for a satellite receiving system installation. In short, none of us want to see a dish sailing off a mount, or worse yet, off a roof during a high wind. Therefore, as we select the perfect location for the reflector antenna, we also need to consider the best methods for minimizing wind loading.

To take the easy way out, we can select a mesh or perforated dish in an effort to reduce wind loading. With this type of reflector, we can cut the wind load by 25 percent as compared to solid reflectors. However, wind loading still occurs and, under the right conditions, can cause serious damage to your system.

For that reason, let's take a longer look at wind loading and methods for prevention. We can calculate the wind load of an object through the following equation:

$$D = 1/2 \ C_d PAV^2$$

where:

D = drag force in Newtons
C_d = drag coefficient
A = the frontal characteristics of an object in ft^2
P = the density of fluid stream (1 lbm/ft^3)
V = velocity of fluid stream ft/sec

Since this equation seems rather complex, let's take a closer look at nature and the variables.

In the equation, P equals the density of a fluid stream. For our purposes, we can consider the fluid stream as standard air. At an altitude of zero feet with dry conditions and a temperature of 15° centigrade, standard air has a density of 1.225 kg/m3. At an altitude of 5,000 feet, or the same altitude as the city of Denver, the density reduces by approximately fifteen percent. If we lower the temperature of the air, the density increases. As an example, lowering the temperature from 15° centigrade to zero degrees centigrade increases the density by five percent. Water vapor also affects the density of standard air. If we increase the amount of water vapor in the air, the density of the air decreases.

The drag coefficient, or C_d, of an object changes according to the amount of surface area directly facing the wind. As an example, a ten-foot dish facing directly into the wind can have a drag coefficient as high as 1.5. Adjusting the dish so that it has a 47° look angle into the wind reduces the drag coefficient to 0.9. In addition, yaw, or side-to-side, movement places additional stress and wind resistance on a dish due to asymmetry and lift.

With this knowledge in mind, we can begin to see whether a dish mounted on a pole will have stability in a windstorm. To accomplish this, we can replace variables in the equation with specific amounts. If we replace P with the worst case scenario where zero percent humidity exists at zero degrees centigrade, then we have:

$$D = 1/2C_d PAV^2 \quad \text{and} \quad D = 1/2C_d^{(1.3)}AV^2$$

From there, we can convert A, or the frontal characteristic of the object, to:

$$A = pd^2/4$$

where:

(d) equals the diameter of the dish

By making these changes, we have converted the original equation into:

$$D = 1/2Cd^{(1.3)} (pd^2/4) V^2$$

Reducing the equation again, we have:

$$D = (0.002133) * C_d * d^2(ft) * V^2(mph)$$

Given this equation, we can replace the drag coefficient, the diameter of the reflector, and the velocity with actual numbers. As an example, the drag force for a 10' diameter reflector with a drag coefficient of 1.5 during 100-mph winds equals:

$D = (0.002133) (1.5) 10^2 \times 100^2$ or
$D = (0.002133) \times 100 \times 10,000$ or
$D = (0.00320) \times 1,000,000$ or
$D = 3200$ lbs.

If a length of steel pipe supports the dish while extending from a concrete base, the maximum stress occurs at the point where the pipe enters the concrete. Since we have the combination of drag force and yaw movement, it becomes crucial to select the proper diameter and thickness for the ground pole. If the installation involves a roof mount, then the same considerations occur along with the use of additional supports for the pole.

Steel pipe arrives in categories of thickness called schedules that affect the inside diameter and the durability of the pipe. As an example, a 3½-inch diameter schedule 80 steel pipe has an outside diameter of four inches and an inside diameter of 3.364 inches. In comparison, a 3½-inch diameter schedule 40 steel pipe has an outside diameter of four inches and an inside diameter of 3.548 inches. Because of the added wall thickness, schedule 80 steel pipe has one-third more the strength of schedule 40 steel pipe.

Chart 4.2 shows the maximum lengths of different diameters of schedule 40 steel pipe when used with different sized reflectors. The maximum length of the pipe in each category varies in accordance with the drag force of the dish and is calculated using a 100-mph wind and a 28° look angle. The 28° degree look angle stands as the minimum look angle and will provide the lowest drag coefficient. The use of a mesh dish for each of the measurements reduces the drag factor by thirty percent. Filling the pipe with concrete increases the ground strength by thirty percent.

Dish Diameter	Drag	3 1/2" Pole	4.0" Pole	4 1/2" Pole	5.0" Pole
6 feet	464	12.19	19.34	N/A	N/A
8 feet	825	6.85	10.89	16.44	N/A
10 feet	1,290	4.39	6.95	10.51	22.04
12 feet	2,180	N/A	5.35*	8.087*	16.95*

Chart 4.2. Ground pole lengths. Pipe outer diameter in inches (schedule 40) (20° Look Angle, 80 mph wind, mesh dish).

Installing the Ground Pole

Figure 4.9 shows a typical ground pole installation. The freestanding ground pole should have a large enough base to contain the wind forces applied to the reflector and the pole. As a rule, add one foot of pipe for every five feet that extends above ground. The hole should have a width four times the diameter of the pole for heights below ten feet. For ground poles that extend above ten feet, the width should increase to five or six times the diameter of the pole.

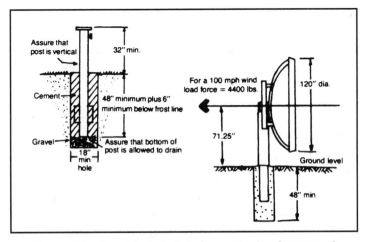

Figure 4.9. Groundpole installation methods. Courtesy of Channelmaster, Inc.

Gravel covers approximately one inch of the hole before the addition of the concrete because of the need for drainage. As the illustration shows, we also drill a hole through the bottom of the ground pole and add a short piece of pipe to prevent the ground pole from turning in the hole. Before pouring the concrete, stake the pole into a vertical position using cables and turnbuckles. Through the use of the turnbuckles and a level placed at different locations around the pole, we can ensure that the pole has a perfect vertical orientation.

In addition, we should also dig the trench used for the cabling that attaches to the feedhorn, the LNB, and the actuator. Although most manufacturers supply direct burial cable for this purpose, the use of plastic conduit ensures the durability of the cable and eases cable replacement when necessary. The cable can extend up from the conduit and fasten to the pole during the final installation of the dish hardware and electronics.

As mentioned, we can increase the strength of the pipe by filling it with concrete. However, before filling the pipe with concrete, determine the maximum level for the concrete. If we fill the pipe completely to the top with concrete then it becomes extremely difficult to insert a bolt through the pipe for the mount. In addition, check the installation to ensure that the cables run outside the pipe rather than down the center of the post.

Along with the proper installation of the ground pole and the filling of the pole with concrete, always attempt to shelter the installation as much as possible from direct winds. Placing the system next to a structure distributes the destructive wind forces over a greater area. If possible, when mounting the pole directly adjacent to a structure, bolt the pole completely through the structure wall and use a backplate on the inside. The backplate should fit adjacent to as many wall supports as possible.

Guy wires can also support the pole. When attaching guy wires to the pole, always use stainless steel cables that will not rust and cable diameters of a 1/4" minimum. Tie the steel cable to the pole through heavy-duty turnbuckles. As the cables loosen over a period of time, the turnbuckle provides a method for tightening the cable to its original tension.

Pouring a Pad

In some cases, the hardness of the ground, a high water table, or the size of the dish antenna may prevent the installation of a ground pole. Any dish antenna that has a diameter larger than 12 feet requires more support than that provided by a ground pole. In those instances, we can pour a concrete pad as shown in *Figure 4.10*. To prepare for the construction of the pad, we excavate a shallow area the size of the pad and build a wooden form around the area. The length and width of the pad should equal at least half the reflector diameter.

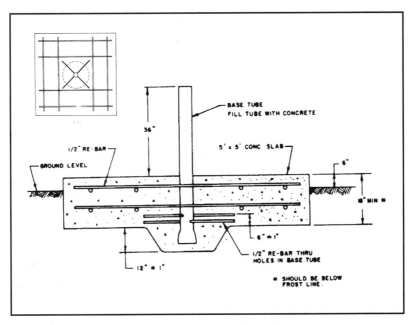

Figure 4.10. Installation of a pad mount.

Then, we add a layer of gravel for drainage and cover the pad area with a wire mesh for additional strength. As with the ground pole installation, we should also accommodate the cabling used to power the LNB, feedhorn, and actuator before pouring the concrete. The concrete used for the pad should have a thickness equal to the diameter of the pole used to support the dish if wire mesh is used to strengthen the concrete. Without the use of wire mesh, the thickness should increase to 1½ times the diameter of the pole. In turn, the gravel used to provide a drainage bed for the pad should have a thickness equal to approximately 50% the thickness of the concrete.

As shown in *Figure 4.11*, the pole may mount to the pad in two different ways. The first method is similar to the ground pole method in that the pole extends into the concrete. Rather than using short pipes or solid bars to prevent the pole from twisting within the concrete, the pipes extend the length and width of the pad. In the second method, a pad mount bolts a tripod assembly that holds the support pole onto the pad through the use of concrete anchor bolts. Some installations may utilize a three-point pad mount where three small, individual pads form the concrete foundation for the mount.

Figure 4.11. A pad mount.

Installing a Roof or Wall Mount

When installing a roof or wall mount, always remain aware of wind loading and the type of attachments used to fasten the pole to the roof or wall. As *Chart 4.2* shows, the length of pipe used for the ground pole depends on diameter, thickness, reflector size, and drag force. Because of the additional wind loading that occurs with a wall-mounted or roof-mounted reflectors, the ground pole should have at least a one-quarter inch thick wall and a two-inch diameter. In addition, space support brackets along the pole. Attach the support brackets with heavy-duty U-bolts that, in turn, bolt onto a metal plate. Always fasten additional supports as close to the mount as possible. *Figure 4.12* shows several options for mounting an antenna to the roof.

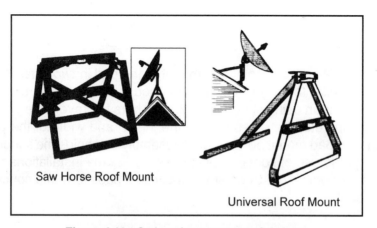

Saw Horse Roof Mount

Universal Roof Mount

Figure 4.12. Options for antenna roof mounts.

For many reasons, most installers prefer not to mount ground poles to chimneys. Usually, the masonry and brick making up the chimney will not support the additional weight presented by wind loading. Moreover, the hot exhaust coming from the chimney can cause the dish to warp or, if the reflector is mounted to close to the chimney opening, catch on fire. If necessity dictates the mounting the ground pole to a chimney, never rely on lag screws drilled into the bricks or masonry. Instead, purchase or fabricate a bracket that fits around the backside of the chimney. The bracket spreads the force over the entire chimney rather than only through the lag bolts. *Figure 4.13* depicts a wall-mounted antenna.

Digging a Trench for the Conduit

Any time that we dig a trench for the cable used between the dish and the dish electronics, we should ensure that the trench will not cut across water, gas, electrical, telephone, or lawn sprinkler lines. Placing the cable within a trench protects the cabling from the elements and from damage caused by small animals, vandals, or accidents. In addition, the placement of cables within a trench adds a finished look to the installation.

The depth of the trench depends on several issues. Some electrical codes require that the cable lie at a particular depth. In addition, we should also place the conduit and cable deep enough so that ordinary gardening will not cut the cable. The use of rigid conduit can help prevent the cutting of cable in this manner. For planning purposes, we should always record the length, route, and depth of the cable run. Do not fill the trench until after testing the system installation.

Most manufacturers offer direct burial cable that encloses all the necessary conductors within a thick plastic sheath. The plastic covering will withstand weather extremes and remains flexible for bends around corners. Despite the ease of installation offered by the direct burial, placing the direct burial, multi-run within conduit provides added protection against damage. Moreover, the conduit provides a method for routing the cable through the trench. If a problem requires the replacement of a transmission cable, we can pull the new cable through the conduit with a fish tape and without redigging the trench.

Types of Conduit

We can purchase conduit in two forms: gray PVC or aluminum conduit. While it may seem cheaper to use the standard white PVC pipe used for water lines, several reasons exist for using the other types. Local electrical codes may not permit the use of the white PVC for carrying any type of conductor. In addition, it does not include any option for the type of sweep needed to pull wiring around a corner.

When installing the conduit at the support pole, always cap the conduit with an electrical weatherhead or a 180° sweep. The use of a weatherhead or sweep prevents moisture from entering the conduit that could either ruin the cable or corrode connectors. To finish weatherproofing the conduit, we can seal the end with weatherproof, flexible putty. A drip loop in the cable as it leaves the conduit also prevents moisture from traveling along the cable and into connectors. In addition to capping the conduit, always use the recommended solvent to glue sections of PVC together. Take care in applying the solvent so that no excess drips onto the cable and damages the insulation. *Figure 4.14* shows a direct burial cable.

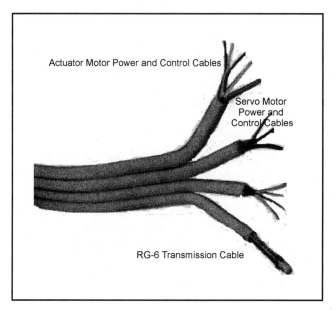

Figure 4.14. Direct burial cable.

When mapping the trench and laying the conduit, ensure that the installation does not include a large number of turns or sweeps and that the conduit has a diameter twice that of the cable. As we pull the cable through the conduit, a large number of turns make it impossible to pull or push the cable past a certain point. In addition, use an electricians wiring snake to pull the cable through the conduit. As you plan the installation, also ensure that the cable length matches the length needed for the installation. Splices in the cable can introduce loss in transmission lines or cause the cable to pull apart.

Grounding the Satellite System

Properly grounding a satellite system has the purpose of limiting voltages and line surges that occur because of near-lightning strikes, unintentional contact with high voltage lines. Moreover, improperly grounded systems can allow the existence of an alternative path around the electrical system of a home. Properly grounded systems ensure the safety of people using the system and the equipment that makes up the system.

Standard 110VAC electrical systems have three wires—black, white, and green or bare copper—attached to the receptacle. While the black wire is the hot connection for the system, the white wire is the neutral lead and the green or bare wire is the ground lead. The white and green or bare wires tie together electrically at the main electrical buss which connects to a ground rod and the neutral connection of the buss. Under normal operating conditions, the white wire carries current as a return back to the source. Again, under normal operating conditions, the green or bare wire carries no electrical current. Under abnormal conditions produced by a fault, the green or bare wire becomes a very low-resistance path and carries current back to the source. The flow of current causes a breaker to trip or a fuse to open and disconnects the faulty equipment from the source.

Electronic equipment such as satellite receivers use the ground connection to establish a baseline voltage as a reference for all internal circuitry. The neutral and ground wires tied together electrically to prevent any increase in eddy currents that increase the electrical potential of the equipment casing. An increase of eddy currents can cause the case to heat to the point where a fire erupts or can establish a shock hazard for anyone touching the case.

Grounding the Transmission Cable

As a rule, always ground the coaxial cable that runs from the LNB to the receiver through a grounding block. In itself, coaxial cable traps any electromagnetic interference and prevents the interference in the outer shield from reaching the inner conductor. Grounding the shield of the coaxial cable reduces any voltages picked up by the outer shield to zero potential and prevents any portion of the stray signal voltage from reaching the inner conductor.

In addition, the coaxial cable ground dissipates static electricity. Because of the sensitivity of the LNB, the dissipation of static electricity is an important factor for a satellite receiving station. An electrical charge can build up on the dish and the coaxial cable until the charge becomes powerful enough to jump across an air space to ground. When this occurs, the static charge will follow a path that leads through either the electronics found within the receiver or in the LNB. As a result, the LNB is damaged beyond repair.

Grounding the transmission cable of a satellite receiving systems also protects the system from lightning strikes. Rather than carry the voltage from a lightning strike to ground, separately grounding the dish and the cable allows the dish antenna to repel the strike. With lightning, large static charges of negative energy build on the base of the cloud. Because the earth holds a positive charge, the negatively charge electrons from the lightning flow to the ground through an air space. The air becomes moist as the charges build and attempt to balance.

As the negatively charged lightning strike continues towards the earth, a positive ground charge travels up through the dish. With this, the dish acts as a lightning rod. The positive ground charge travels up through the dish and neutralizes the downward negative strike. With the two opposing electrical charges neutralized within a few thousandths of a second, the negative charges empty from the cloud and the positive charge at the ground dissipates.

The effectiveness of any lightning rod system depends on the capability of the system to spread a positive charge over the entire protected area. Because of this, simply grounding only the satellite system rarely provides the necessary protection. A lightning strike could dissipate directly through the dish. Most contractors recommend mounting a Y every 20 feet along the highest edge of the roof. Always connect the rods to each other at the highest point on the roof and work with the connections in a consistent downhill manner while connecting everything that rises above the roof to the rods. The failure to keep all points on the roof and all electronic equipment grounded to the same reference can create a situation where everything exists at different ground potentials. All electrical codes require the unified bonding of grounds. Without unified bonding, or the connection of all grounds to the same reference, current will flow from the point that has the higher ground potential.

Protecting Your System from Surges

Even with the possibilities for lightning strikes, the greatest electrical danger for electronic equipment extends from voltage surges. Near-miss lightning strikes can induce voltage surges in house wiring. Because of the massive amount of energy contained within lightning, a near-miss can cause an electromagnetic pulse that induces current within any wire found within the magnetic field. Although a voltage spike may occur within a fraction of a second, the sharp rise in current destroys circuits designed for minimal current and voltage shifts. Surge protection equipment contains metal oxide varistors, or MOVs, that offer infinite resistance at one voltage and decreasing resistance at higher voltages. In low-cost surge protectors, an MOV connects between the hot and neutral leads and protects against voltage spikes generated by electrical devices such as motors. More expensive surge protectors that also protect against voltage spikes generated by lightning strikes connect one MOV between ground and hot and another MOV between neutral and ground. As a result, the surge protector can clamp a 300-volt surge down to 15 to 30 volts. Other surge protectors combine MOVs with triacs and varistors. Regardless of the configuration, surge protectors are permanently damaged after a voltage surge.

Selecting an Antenna Positioning System

One of the keys to receiving the maximum enjoyment from a C-band TVRO system rests with selecting the best positioning system. The combination of positioner and actuator should provide a reliable, convenient, cost-effective means for moving the dish across the Clarke Belt. Lack of reliability causes the loss of viewing time and introduces service expenses. While we could rely on a hand crank to move the dish across the Clarke Belt, convenience points towards the use of an electronic control system and a motor.

An antenna positioning systems combines machinery and electronics. We can divide the system into four parts. Those are:

•   Actuator

•   Motor drive circuit

•   Antenna position indicator, and

•   Memory and control circuit

An actuator consists of a motor, a set of gears, and a power jack that extends or retracts. One end of the actuator attaches to the reflector while the other end pivots on a fixed antenna support. As the jack extends or retracts, the antenna rotates on its axis. A good actuator should have more than sufficient thrust and strength along with travel limits and weather protection.

Thrust involves the power needed to move the reflector along its axis. An actuator with adequate thrust can move the antenna despite high wind conditions or loads created by snow or ice. We can define the strength of the actuator through its structural integrity and the mounting brackets that secure the device to the dish and mount. The actuator should have sufficient strength to prevent the antenna and actuator from coming loose during extreme weather conditions. Mechanical and electronic travel limits ensure that the jack will not extend past its physical capabilities and that the dish will not bind against the mount. Weather protection prevents water from entering any portion of the actuator.

Actuators

An actuator consists of a motor and gear assembly and mechanically drives the dish antenna East or West across the satellite arc. Actuators may take the form of linear drive designs or horizon-to-horizon drive designs and feature different types of gear or screw assemblies. Depicted in *Figure 4.15*, a linear drive actuator utilizes a telescoping arm that moves within a fixed tube. While one end of the linear actuator attaches to the mount, the other end attaches to bracket located on the dish. A linear actuator can only sweep across approximately 100° of the arc created by the Clarke Belt. Horizon-to-horizon actuators place the entire gear mechanism at the fulcrum of the dish and can track satellites between the East and West horizons. *Figure 4.16* shows another type of horizon-to-horizon actuator. As the figure shows, the mount and actuator combine into one mechanism. With this, the horizon-to-horizon mount offers easier installation and less susceptibility to water damage than that seen with linear actuators.

The capability of any type of actuator to move the dish antenna depends on the weight of the antenna and the wind loading conditions. Under high wind conditions, a solid dish may couple anywhere from 400 to 1800 pounds of force with its actual weight. Although perforated or mesh dishes remain less susceptible to wind loading, certain conditions can create comparable forces. Wind loading affects both the mount and the actuator.

Acme Thread and Ball Screw Actuators

An Acme thread actuator uses a threaded screw shaft that travels within a threaded collar. Acme thread actuators can move up to 800 pounds. Instead of relying on a threaded collar, the ball screw actuator uses ball bearings as the friction element.

ITEM	PART NO.	DESCRIPTION
1	7821646	KIT, LIFT ASM (18.0 STROKE-3/5)
2	5703357	SCREW, THD. FORM. (3/8-16 x .85)
3	7820354	KIT, MOTOR (36V DC)
4	7820310	HOUSING MOTOR (36V DC)
5	7821836	SHAFT, INTERMEDIATE
6	120217	LOCKWASHER (#10)
7	7821096	GEAR, INTERMEDIATE
8	7820318	GASKET, GEARBOX
9	7821097	GEAR, OUTPUT
10	5703499	PIN, GROOVE (TYPE F)
11	7821826	HOUSING, GEAR
12	7821811	GROMMET, RUBBER (MTR & GRBX)
13	7821181	SCREW, THD. FORM. (6-32 x 1.00)
14	7821798	STRIP TERMINAL
15	7821797	SCREW, TERMINAL
16	7820318	GASKET, COVER
17	7821822	SCREW, BRACKET
18	7821178	POTENTIOMETER (10 TURNS) 10K ohm
19	7821831	BRACKET, POTENTIOMETER
20	7821838	LOCKWASHER
21	7821837	NUT
22	7821176	GEAR, OUTPUT
23	7821170	GEAR, INTERMEDIATE
24	7821169	GEAR, PINION
25	7821824	PIN, GEAR
26	7820309	COVER, HOUSING
27	7821179	SCREW, FIL. HD. (10-24 x 2.625)
28	316009A	CABLE CONNECTOR
29	3210025	RUBBER WIPER
30	1320004	BELLOWS
31	2230094	BOOT

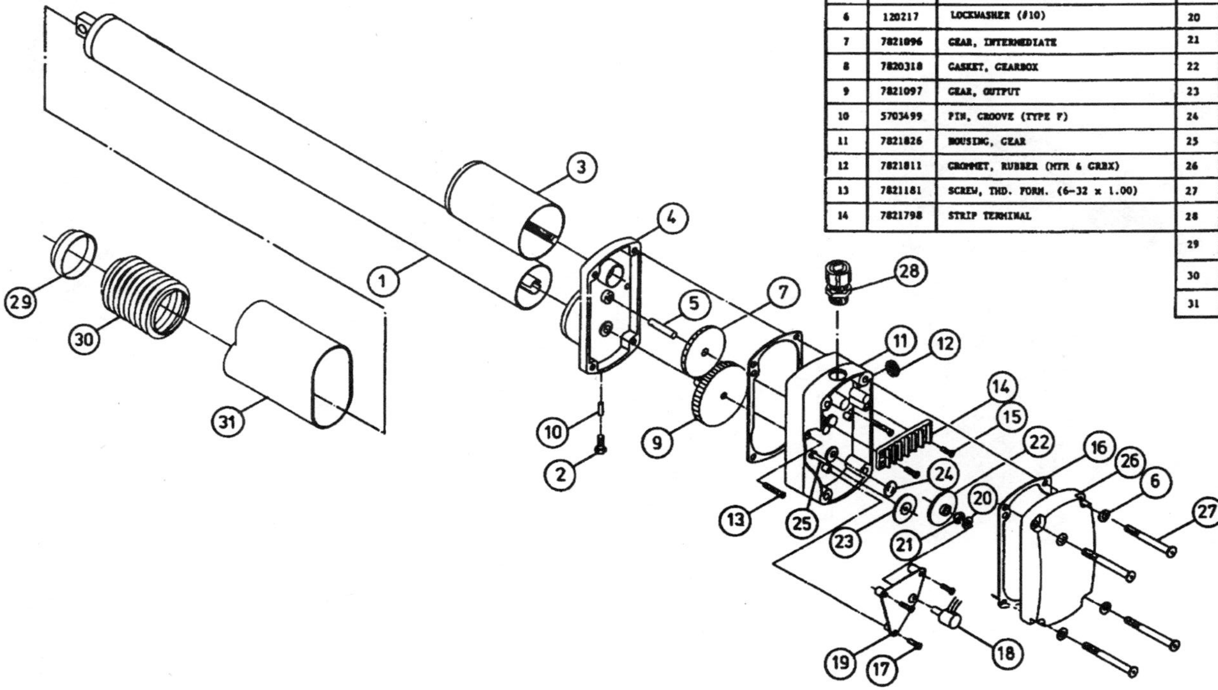

Figure 4.15. Exploded drawing of a linear drive actuator.

119

Because of the use of ball bearings, ball screw actuators have less loading than acme thread actuators and allow the motor to transfer more energy into moving the dish. With the ball bearing assembly providing smoother actuator movement and a greater transfer of energy, the ball screw actuator has greater reliability under changing environmental conditions.

Figure 4.16. A typical horizon-to-horizon mount.

Feedback Elements

Every actuator has mechanical limits that establish East and West boundaries for the satellite arc. Linear actuators have limitations in that the arm assembly should never push or pull a dish through an angle of less than 30°. Any angle of 30° or less between the arm and the back of the dish places extreme lateral force on the arm and could bend the internal mechanism. To guard against this problem, actuator designs may have a variety of feedback elements that allow communication between the actuator and the receiver. Those feedback elements may consist of optical sensors, reed switches, Hall effect sensors, or potentiometers.

Optical Sensors

Optical sensors rely on the light emitted by a light-emitting diode, or LED, and a phototransistor. During operation, the phototransistor uses the interruption of the LED emission by the motion of the motor to generate a feedback pulse. The pulse travels to the actuator controller and notifies the controller about the position of the dish. Every 20 to 40 pulses equal one inch of travel by the actuator arm.

Reed Switches

Reed switches generate a pulse when exposed to a changing magnetic field. During the setup of the actuator, either mechanical limit is designated as zero while the other mechanical limit functions as the maximum count. With the magnet attached to the motor through a set of gears, each rotation of the magnet causes the reed switch to close and then open once for every rotation. As a result, the reed switch generates a pulse for each rotation of the motor. The pulses travel to the controller circuitry and create a continuous count of the actuator movement. Every 20 to 40 pulses equal one inch of travel for the arm assembly.

Hall Effect Sensors

Hall effect sensors use a semiconductor element that has sensitivity to magnetism to generate a pulse for the actuator controller. Referring to *Figure 4.17*, a rotating magnet moves across the face of the semiconductor as the actuator moves the dish. The resulting pulses work as counters for the controller. With Hall effect sensors, every 20 to 40 pulses equals one inch of travel for the actuator arm.

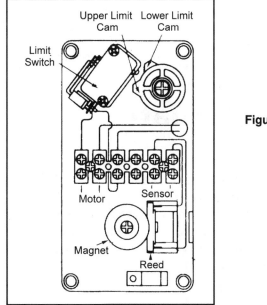

Figure 4.17. Actuator using a hall effect sensor.

Potentiometers

Older actuators rely on a ten turn, 10,000 ohm potentiometer to generate the feedback signal for the actuator controller. Because the potentiometer attaches to a set of gears that, in turn, attach to the motor, any motor movement turns the potentiometer shaft. As a result, the motor movement varies the resistance of the potentiometer. The changing resistance works as a control pulse for the actuator controller. *Figure 4.18* shows a typical potentiometer feedback element.

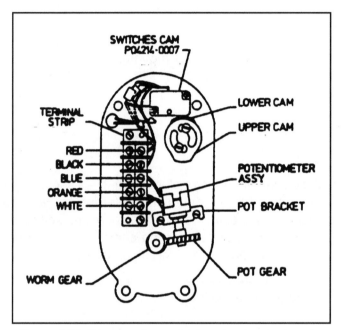

Figure 4.18. Diagram of an actuator using a potentiometer feedback element.

Attaching the Actuator

Although attaching an actuator may seem as simple as fastening two bolts and connecting four cables, the proper installation of the actuator calls for several steps. As we attach the actuator to the mount and reflector, we should always ensure that the arm does not bind. The actuator must attach to the mount and reflector thorough ball joints so that lateral movement can occur freely. Any side pressure on the actuator can cause the inner tube to bend and finally seize.

In addition, the attached actuator arm should form a 30° angle with respect to the rear of the reflector and at all points as the antenna sweeps across the arc. If we decrease the angle, the actuator loses leverage and the electric motor found within the actuator works harder than specified. When installing the actuator, ensure that the device has fully retracted. Position the mounting clamp so that the reflector can aim several degrees past the lowest satellite on the horizon.

The length of the actuator can vary from 12 inches to more than 48 inches. Larger reflectors that have increased weight and wind loading require longer arms for added leverage. With this type of installation, attach the actuator arm parallel to the plane of rotation. As before, attaching the actuator in this way minimizes the lateral forces that can bend the inner tube.

Setting the Mechanical and Electronic Limits

Adjustable limit switches found within the actuator prevent the actuator from overextending or retracting to a point where the dish and actuator arm become damaged. Accidentally overextending the arm can cause the actuator to freeze in the extended position or—at worst—bend the arm. In addition to the mechanical limit switches, the positioner control also includes electronic limits set during the installation. During the installation of the actuator, position the dish just beyond the lowest satellites in the arc while setting the limit switches and electronic limits.

Actuator Cables

Every actuator requires voltages needed to move the dish and to control the position of the dish. The heavier guage cables carry the 36 VDC that powers the actuator motor. Smaller gauge cables carry the counting pulses between the actuator controller and the actuator sensor. In each case, the length of the cable run determines the size of the cable used for the installation. Any installations with cable runs of 400 feet or more will use 12 gauge cables for the actuator motor while 14-gauge cables work well for cable runs of less than 400 feet. Since controlling the actuator depends on interaction between the controller and some type of servo within the actuator, the controller cables vary from 20 gauge for shorter runs to 18 gauge for cable runs that extend beyond 400 feet.

Summary

Chapter four makes the transition from the theoretical base built within the first three chapters to practical advice about installing the major hardware sections of a C-band satellite receiving system. The chapter begins the transition with instructions for assembling different types of reflector antennas. From there, the chapter compares different types of mounts used for dish antennas. Specifically, the chapter provides information about AZ-EL, polar tracking, and horizon-to-horizon mounts. In addition, the chapter provides instruction for mounting the ground pole and for pouring a cement pad.

The discussion about installing the reflector, mount, and pole also includes valuable information about wind loading. In addition, the chapter guides the reader through information about properly grounding the system at the pole and at the transmission cable. While discussing grounding techniques, the chapter also addresses surge protection. The chapter continues with a description of different types of actuators and shows the type of cables used for actuator control. The discussion of actuators also covers common faults that cause actuator failure and possible damage to the dish antenna.

CHAPTER

Installing the
Feed and Amplifier

Chapter five continues with the practical installation advice given in chapter four. However, the discussion moves from the emphasis on hardware in chapter three to an emphasis on the electronic portion of the dish antenna. The first portion of the chapter considers the characteristics of a good antenna feed. Along with this consideration, the chapter also describes servo motors used to adjust the feed antenna for the proper polarity. The section about antenna feeds concludes with instructions for installing and aligning the antenna feed.

Chapter five uses many of the terms first defined in chapter one during the discussion about low noise amplifiers and low noise block downconverters. The chapter provides a detailed look at LNBs with a discussion about the internal circuitry of LNBs and the affects of noise and temperature on amplifier operation. The chapter also provides information about installing an LNB on the system. As the chapter concludes, it offers additional specifications about the transmission cables and connectors.

The Antenna Feed

Everyone recognizes a satellite receiving system because of the presence of the dish antenna. Yet, the dish antenna merely serves as a reflector that concentrates the signal received from the satellite to the focal point. Once the signal arrives at the focal point, the feedhorn collects and passes the signal onto a low noise amplifier, low noise converter or a low noise block downconverter. Many direct broadcast systems utilize a one-piece low-noise feed that combines the feedhorn assembly and the amplifier assembly. Feedhorns used in C-band and Ku-band systems bolt directly to the amplifier and has the appearance shown in *Figure 5.1*.

Figure 5.1. C-Ku-Band feedhorn. Courtesy of Chaparral System,Inc.

An antenna feed couples the incoming RF energy reflected from the dish into the LNB. Given this task, a feed can affect the performance of the entire system in several ways including:

- The thermal noise temperature — As we know, the feed should see as much of the dish surface as possible. Systems with either prime focus or offset feeds should not have overspill of the feed reception pattern. Otherwise, thermal noise emitted by the ground will increase the antenna equivalent thermal noise temperature.

- Efficiency — An optimum feed design will provide maximum antenna efficiency with zero attenuation of the illuminated part of the dish but infinite attenuation for angles beyond the dish. In practice, feeds can only approach the uniform illumination law.

• Bandwidth — Inverse taper illumination given through the feed can provide a main lobe that has a narrower beamwidth.

Figures 5.2 and *5.3* illustrate the uniform illumination and aperture edge taper laws while *Figure 5.4* shows feed overspill and inadequate feed illumination.

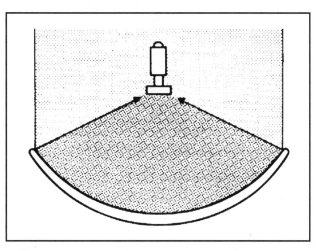

Figure 5.2. Uniform illumination of the reflector by the feed.

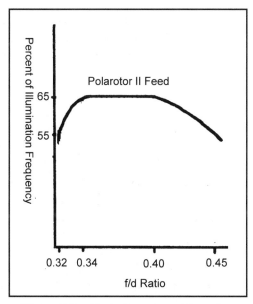

Figure 5.3. Illustration of aperture edge taper laws. Courtesy of Chaparral.

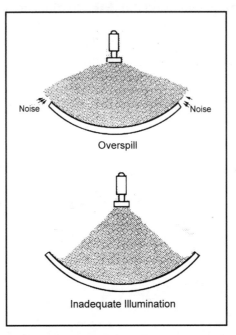

Figure 5.4. Illustration of overspill and inadequate feed illumination.

The correct positioning of the feed in relationship to the reflective surface of the dish is crucial for the performance of the satellite reception system. If we view the feed as a flashlight placed above the curvature of the dish antenna, the light emitted by the flashlight should illuminate the majority of the dish surface. In terms of the feed, the correct positioning of the feed illuminates the center portions of the reflective surface more strongly than the outer portions. As a result, the feed collects the microwave signal reflected from the center of the dish and misses the signal reflected from the edges of the dish antenna.

From the description of beamwidth and sidelobes, we know that the signals reflected from the edges of the dish contain a combination of terrestrial and adjacent satellite interference. While a feed that illuminates the entire dish would improve gain, it would also capture some of the interfering signals. However, a feed positioned so that it illuminates only a small portion of the dish antenna center will lose gain and miss much of the signal reflected from the dish. Because of this, manufacturers specify the correct focal distance for particular types of dishes.

Although most of us associate the term "antenna" with the dish, a small probe found either in the mouth of the feed or within the amplifier functions as the actual antenna in a satellite receiving system. Illustrated in *Figure 5.5*, this small antenna features precise tuning and has correct dimensions and position to convert microwave signals to usable signals.

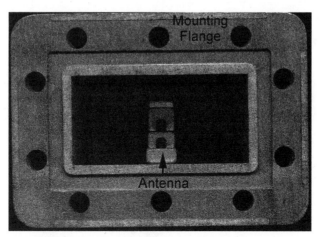

Figure 5.5. Antenna probe found inside the LNB.

Focal Length and F/D Ratio

On any specific parabolic reflector antenna, one best or optimum location exists for the placement of the feedhorn. As mentioned in chapter two, the *focal length* of a dish measures the distance from the dish surface to the focal point. In more specific terms, the focal point of the dish must land one-quarter inch inside the throat of the feed. Most manufacturers provide the exact focal length needed for the reflector antenna.

Without that information, we can find the focal length of the reflector through the use of the following formula:

$$f = (CD)^2 / 16(FE)$$

where:

f = the focal length
CD represents the diameter of the reflector and
FE represents the depth of the reflector

The distance for the focal length always measures from the center of the dish to one-quarter inch inside the feed. *Figure 5.6A* illustrates the focal point, focal length, depth, and diameter of a reflector antenna, while *Figure 5.6B* shows a technician measuring the focal length of a new installation.

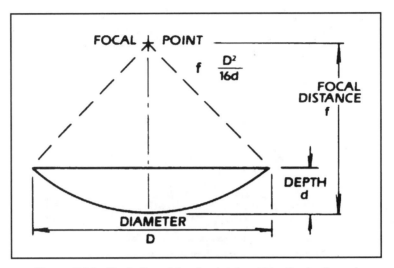

FOCAL ★ POINT

$f \quad \dfrac{D^2}{16d}$

FOCAL DISTANCE
f

DEPTH
d

DIAMETER
D

Figure 5.6A. Illustration of focal point, focal length, depth, and diameter of a reflector antenna.

Figure 5.6B. Measuring the focal length.

f/D Ratio

We can classify dish antennas according to a focal length to diameter ratio, or f/D. Low f/D ratios indicate a shorter distance between the feed assembly and the reflective surface of a dish antenna that has a given size while high ratios indicate a longer distance. The distance between the feed and the reflective surface determines the ability of the feed to illuminate the reflective surface.

Through the use of the f/D ratio, we can establish whether a dish is deep, average, or shallow. A deep dish will have a f/D ratio in the 0.25 to 0.35 range while a shallow dish will have a ratio in the 0.4 to 0.5 range. Deep parabolic dish antennas have a smaller f/D ratio and, as a result, have a narrow beamwidth, smaller side lobes, and lower noise temperatures because the feed assembly sits closer to the reflective surface. However, shallow reflector antennas have lower side lobes, provide better noise temperature performance, and better efficiency. *Figure 5.7* illustrates the differences between deep and shallow dish antennas.

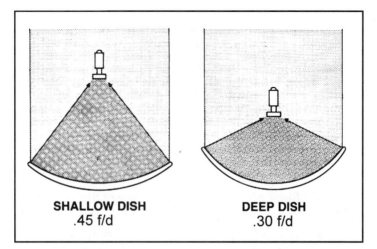

SHALLOW DISH
.45 f/d

DEEP DISH
.30 f/d

Figure 5.7. Difference between deep and shallow reflector antennas.

To find the f/D ratio, we can either consult the manufacturer's guidelines or use the following formula:

$$f/D = f/CD$$

where:
f represents the focal length of the reflector antenna and
CD represents the diameter of the reflector antenna.

Scalar Feedhorns

Scalar feedhorns remain as the most common type of feedhorn manufactured today. Pictured in *Figure 5.8*, a *scalar feed* relies on a series of concentric rings to gather and direct the signal from the outer edges of the focal point to the center of the feed. Most of the signal picked up by the feedhorn occurs because of reflection from the inner portion of the antenna surface area, or approximately 70 percent of the dish surface.

Figure 5.8. Scalar feed.

The design of the scalar feedhorn maximizes the view of the dish and attenuates 8 to 22 decibels of the signal received from the edges of the dish. In addition, the attenuation provided by the scalar feed prevents molecular noise generated by the earth from interfering with the desired signal. The molecular motion within the Earth generates random noise that covers almost all the electromagnetic spectrum used for the transmission of satellite signals. Attenuation of this noise becomes necessary because the noise has a signal strength several times stronger than the strength of the satellite signals. With scalar feeds, attenuation takes the form of an illumination taper that sharply reduces the reception of Earth noise found past the rim of the antenna.

Polarization

Going back to chapter one, all radio frequency waves, including satellite television signals, have polarization. In brief review, designers use the property of satellite signals to improve the spectrum efficiency in satellite bands. The use of different polarities allows the transmission of more signals from the satellite without the risk of interference from one channel adjacent to another. Because of the different polarization schemes, the feed used within a satellite reception system must have the ability to change polarization formats. Nearly all satellite operators use opposite polarizations to provide more transponders within a limited bandwidth.

Satellite television applications may use either circular polarization or linear polarization. With the circular polarization, signals arrive as either right-hand polarized or left-hand polarized. With linear polarization, the signals arrive as either horizontally polarized or vertically polarized. Most satellite applications use linear polarized signals; in turn, most TVRO systems rely on linear polarized feedhorns. While optimized for linear polarized signals, this type of feed can receive circular polarized signals with an inherent 3 dB loss.

Linear Polarized Feeds

When properly installed, a linear polarized feed is optimized for the reception of one polarization of the satellite signal with high attenuation of the opposite polarization. Because satellites carry both vertically and horizontally polarized signals, the receiving system must employ some type of strategy for receiving both types of signals. Although we could manually rotate the feed and amplifier assembly through 90 degrees, modern systems use a small motor to rotate only the antenna probe found inside the feed rather than the entire assembly.

Servo Motors

A servo motor turns the resonant probe found in the waveguide throat. During operation, the servo motor relies on a dc voltage supply, a ground connection, and a pulse width voltage that switches the supply across the motor. As the pulse width supply switches the supply, the probe moves from one polarization format to another. Each pulse is applied for a fixed amount of time and determines how far the probe will move.

Achieving the Proper Polarization with a Linear Feed

Although early systems rotated the entire feed, almost all satellite reception systems seen today use either a mechanical, magnetic, or pin diode method for changing the polarity of the antenna located within the feed. Shown in *Figure 5.9*, the *mechanical* method uses either a servo or a small dc motor to rotate a small metal hook found within the circular portion of the waveguide between the horizontal and vertical polarities. The hook features a probe section that detects signals at either polarization. At the other end of the hook, a straight stem passes through plastic dielectric material and radiates the signal into the rectangular portion of the waveguide. As the signal travels from the probe into the rectangular portion of the waveguide, it retains the original polarity of the transmitted signal.

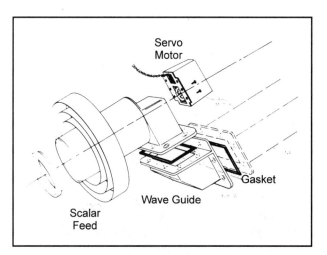

Figure 5.9. Using a servo motor to change polarity.
Courtesy of Chapparal Systems, Inc.

Orthogonal Mode Transducers

A transducer is an electronic device that converts one type of energy into another. Some commercial C-band satellite reception systems use orthogonal mode transducers to simultaneously detect vertically and horizontally polarized signals. In addition, dual band probes use orthogonal mode transducers to switch between C-band signals and Ku-band signals.

Dual Polarization Antenna Feeds

Other approaches to the polarization problem involve the use of dual polarization antenna feeds. During operation, the probe designated for horizontally polarized signals detects only those signals while the probe designated for vertically polarized signals receives only those signals. Instead of using a waveguide section, some dual feed designs rely on two independent probes and two low noise amplifiers built into a single housing.

Magnetic Polarization

Magnetic polarization used for this type of feed does not rely on moving parts. Instead, polarity changes as incoming microwave signals intersect a magnetic field. With this method, a magnetic selector consisting of a ferrite rod fastened within the circular waveguide by a dielectric support produces a magnetic field. Current passes through the a large coil wound around the outside of the waveguide and produces the magnetic field. When the flow of electric current changes direction, the orientation of the magnetic field changes and affects the polarity of the microwave signal.

Pin Diode Polarization

Another approach involves the use of two probes mounted at right angles from one another. Diode switches select the appropriate probe. *Pin diode polarization* relies on the electronic switching between two probes located within the circular portion of the waveguide. While one probe detects horizontally polarized signals, the other detects vertically polarized signals. A voltage carried by a cable feeding from the low noise amplifier changes during operation and switches between the two probes.

As the voltage level changes between two distinct levels, the pin diodes contained within the probes turn off and on. In comparison with the mechanical and methods of polarization, pin diode polarization does not use any type of skew adjustment. The use of a skew adjustment allows the system to match the polarity as it continuously changes between horizontal and vertical.

Circular Polarization

The mechanical, magnetic, and pin diode polarization methods refer to linearly polarized satellite signals. However, some satellites utilize circular polarization rather than the more common linear polarization. A scalar feed designed to receive circularly polarized signals has a rectangular piece of teflon dielectric inserted within the circular waveguide. The teflon dielectric sits at a 45° angle in relationship to an internal probe. During operation, the dielectric delays the circularly polarized signal and converts the signal into a linearly polarized format. From there, the feed can differentiate between the vertical and horizontal signals.

Installing the Feed

The assembly and proper alignment of the feedhorn exist as absolutely critical operations for the performance of the satellite receiving system. To emphasize this point, a properly installed feed system can improve reception by as much as two decibels. Of course, we can also lose two or more decibels of gain with an improperly installed feed. Rather than depending on the manufacturer's design to ensure that the feed aligns exactly with the center of the reflector antenna, always perform three measurements to verify the accuracy of the feed alignment. The measurements are the focal length, feed center, and feed plane.

Two different methods exist for installing the feedhorn to the reflector antenna. Illustrated in *Figures 5.10* and *5.11*, the feedhorn may mount to the reflector through a tripod arrangement or through a button hook mount. The tripod support for the feedhorn fastens at three equidistant locations at the dish and mounting holes found at the feedhorn. In contrast, the button hook support extends through the center plate of the reflector. The feedhorn fastens to a mounting plate found at the top of the button hook support.

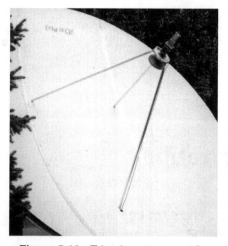

Figure 5.10. Tripod arrangement for mounting a feed assembly.

Figure 5.11. A button hook feed mount.

Finding the Feed Center

We can use the illustration of a billiard ball and a pool table in *Figure 5.12* to show the importance of exactly centering the feed on the axis and parallel to the face of the reflector. If we set up a billiard ball at one end of the table and attempt to bounce it off the bank at the opposite end with the hope of returning it to the original spot, we need to hit the ball at a precise 90° angle. If we hit the ball at anything less or more than 90°, the ball misses the original spot by several inches to either side.

The feedhorn and parabolic reflector have the same relationship as seen with the illustration of the billiard ball and pool table. To ensure that the signal reflected from the dish strikes the feedhorn at the proper angle, measure from three equidistant points on the perimeter of the reflector to the edges of the feedhorn. Add each of the measurements and then divide by three. The average given by this calculation provides the correct measurement for centering the feed with the reflector.

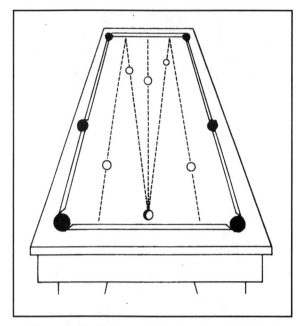

Figure 5.12. Schematic drawing for a low-noise amplifier.

Moving to *Figure 5.13A*, carefully adjust the feed mount in the correct direction so that all three measurements equal one another. If slightly bending the feed mount does not yield the correct measurements, install three guy wires with turnbuckles. While the guy wires hold the feed in place, adjusting the turnbuckles allows the fine tuning of the measurement.

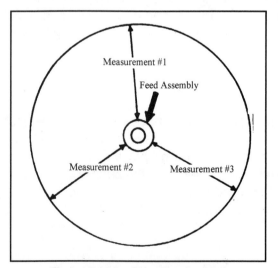

Figure 5.13A. Centering the feed.

After ensuring that the feed remains centered with respect to the reflector, determine if the face of the feed lies parallel to the face of the dish and that the feed aligns with the polar axis as shown in *Figure 5.13B*. Measure from three equidistant locations on the perimeter of the feedhorn to the center of the reflector. Each measurement should equal the focal length of the antenna. If we find unequal measurements, the feedhorn has skewed slightly to one side. With this, the feedhorn will unevenly illuminate the reflector and reduce the signal gain. To maintain the correct feedhorn alignment, gently twist the feedhorn in the appropriate direction.

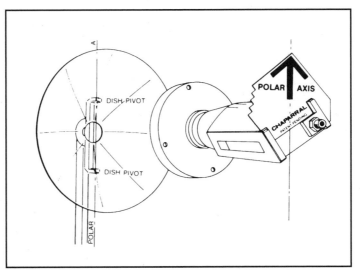

Figure 5.13B. Aligning the feed with the polar axis of the dish.

Polarity Switching and Skew

In chapter two, we found that the probe located within the waveguide of a polarizer must have the same orientation as the incoming signal. That is, the reception of a horizontally-polarized signal requires the horizontal orientation of the probe. The same relationship occurs with the reception of vertically-polarized signals. All this sounds simple until we realize that a reflector attached to either a polar mount or a horizon-to-horizon mount changes angles slightly as it sweeps across the Clarke Belt.

We can offset the angle through the automatic setting of the polarization offset. To accomplish this, we correctly align the probe with the system set to the Southern-most satellite. Most receivers use a voltage-switched polarity feed. Aligning the probe involves switching the probe to either horizontal or vertical polarity and then physically rotating the polarizer assembly to a horizontal or vertical position before tightening the feed mounting bolts.

Proper alignment of the feed becomes even more important when we consider that the servo motor controlling the probe has mechanical limits that provide only a 270 degree range of motion. If the limits of the motor do not extend past the orientation of the vertically and horizontally polarized signals, the probe cannot achieve the alignment needed to receive a strong signal. The probe must have the capability to move 90° between the vertical position and the horizontal position.

The skew adjustment within a polarizer maintains the alignment of the probe with the direction of the polarized signal as the dish sweeps across the Clarke Belt. Modern satellite receivers include an automatic skew adjustment that slightly changes the probe angle as the dish moves. The receivers also contain fine tuning adjustments that allow the customer to manually adjust the skew. Skew adjustments also compensate for satellites that transmit signals along different orientations of the linear polarization format.

Polarizer Cables

The signals that drive the polarizer from horizontal to vertical polarization travel along the 20 guage cables that attach between the receiver and the polarizer. If the receiver attaches to a mechanical polarizer, three cables provide the ground, pulse, and +5 VDC connections for the polarizer. A receiver attached to a ferrite polarizer uses two cables to carry the drive signals.

Because of the chance that electrical interference could drive the polarizer, a grounded aluminum sheath always surrounds the polarizer cables. In addition, the control cables have the minimum gauge and length requirements shown in *Table 5.1*. Those requirements vary with the application. As an example, an installation that features a C/Ku-band dual polarizer will have heavier guage cables.

Maximum Length	Usable Wire Size
25 meters	20 guage
50 meters	18 guage
50 to 100 meters	16 guage

Table 5.1. Polarizer cable specifications.

Amplification and Downconversion

In chapter two, we found that amplifiers increase the gain of a signal. Amplification of the very low power signal received from a satellite occurs once the concentrated signal funnels through the feedhorn. There, the resonant antenna directly couples the signal to the first stage of electronic amplification. Any amplification of the signal also increases the chances for amplifying noise generated from external sources and from within the amplifier circuitry. Manufacturers of amplifiers used for satellite systems use gallium arsenide transistor technology to minimize the noise temperature of the amplifiers and provide "noise figures" for each product. *Figure 5.14* depicts a low noise amplifier.

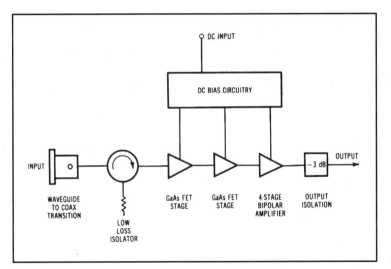

Figure 5.14. Low-noise amplifier.

Single Downconversion

A *single downconverter* translates the incoming signal from the satellite to the final 70 MHz intermediate frequency required for the receiver within a single stage. In most single downconversion systems, the use of a tuner control on the receiver selects a voltage that travels to a voltage controlled oscillator, or VCO, located at the feed. The action of the VCO and associated circuitry produces a frequency 70 MHz higher or lower than the center frequency of the desired satellite channel. Once mixing occurs, a difference frequency of 70 MHz remains.

Dual Downconversion

Dual downconversion systems use two stages usually located in the downconverter to convert the incoming signal from the satellite to the 70 MHz needed by the receiver. As a result, a frequency of 810 MHz functions as the intermediate frequency. Once the consumer selects the desired satellite channel, a VCO heterodynes the center frequency with a signal that has a frequency equal to the center frequency minus the intermediate frequency. After heterodyning occurs in the first stage, the intermediate frequency then mixes with the local oscillator signal to produce the final 70 MHz intermediate frequency.

Block Downcoversion

Block downconversion occurs through the use of a fixed frequency local oscillator that lowers the entire band of satellite frequencies to an intermediate range. Whether working with a C-band signal at 4 GHz or a Ku-band signal at 12 GHz, a low noise converter translates the incoming signals to a range within 950 MHz to 1450 MHz. As a result, the intelligence contained within the entire high frequency satellite bandwidth travels to the satellite receiver. The down conversion of the frequency in the form of a block allows the process to occur without the tuning of the frequencies.

Low Noise Converters

Down conversion of the signals received from the satellite occurs for several reasons. Electronic systems process lower frequency signals much more efficiently than high frequency signals. The captured incoming signal has an electronicmagnetic form. Downconversion of the signal not only lowers the frequency but also changes the electromagnetic signal into an electrical signal. In addition, the high frequency signals experience higher degradation when transferred along a cable.

During the evolution of satellite receiving technologies, manufacturers have employed three different types of downconversion methods called single downconversion, dual downconversion, and block downconversion. Regardless of the method used, downconversion also involves other basic processes and concepts called oscillation, heterodyning, intermediate frequencies, and superheterodyning. With each satellite channel occupying a bandwidth of 18 to 28 MHz, the selection of any one of the channels involves converting the band of higher frequencies to a band of lower frequencies that has the same width. However, the lower band of frequencies centers around a final intermediate frequency.

Low Noise Block Downconverters and Low Noise Block Downconverter Feeds

A *low noise block downconverter* places a low noise amplifier and a low noise converter within the same package. Given this combination, resonant antenna found within the amplifier portion of the package detects the signal reflected by the dish and converts the signal to electrical current, the amplifier increases the gain of the signal by 50 to 65 dB, and the low noise converter downconverts the frequencies of the incoming signal to an intermediate. Because of the use of an amplifier and a converter within the same package, the LNB differs from traditional electronic circuits because of the presence of both electrical and electromagnetic signals. While the use of a resonant cavity type of oscillator defines the internal dimensions as well as the circuit layout and the physical casing for the LNB, the combination of a fixed local oscillator and a mixer found within the low noise converter provides the downconversion.

Low noise block converter feeds take the combination of amplifier and block downconverter a step further by enclosing the LNB within a feedhorn housing. As an example, a dual-band LNBF replaces the waveguide section of the traditional feed with two probes and two LNBs built as a unit within the housing. During operation, each individual probe/LNB set detects the signals of one polarity while the other set detects signals from the other polarity.

Inside the LNB

Referring to *Figure 5.15*, every C-band and Ku-band LNB has a particular shape that occurs because of several reasons. The voltage standing wave ratio, or VSWR, of an LNB measures the effectiveness of collecting the input signal through the division of the power of the input signal by the amount of signal entering the LNB. Because the VSWR must remain lower than 1.3 to 1, LNBs have a specifically shaped waveguide. The shape and dimensions of the waveguide channel precisely channel the C-band or Ku-band microwave signals to the internal antenna probe.

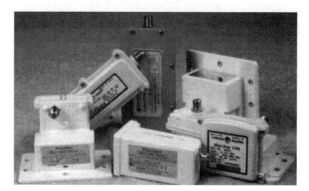

Figure 5.15. A group of LNBs.

Found inside the LNB, the internal antenna probe receives the microwave signals and translates the signals into electrical current. The positioning of the probe within the LNB maximizes the signal reception. Referring to *Figure 5.14* and the photograph of the amplifier circuitry in *Figure 5.16*, the internal antenna probe connects to the electronic circuitry of the LNB which consists of several cascaded GaAsFET transistor stages that precede several more amplifier stages. The combination of the amplifier stages and filters amplify and shape an input signal with an extremely low power of 10^{-14}watt/m^2 so that it appears as a usable signal to the receiver. Bandpass filters within the LNB attenuate any signals that fall outside the designated frequency range of the device.

Figure 5.16. LNB circuitry.

In addition to the amplifier stages, a voltage regulator maintains the consistency of the voltage supply for the LNB. When connected to the TVRO system, an LNB draws approximately 80 to 150 milliamps of current through the IF cable. Most LNBs require a supply voltage that ranges between 15 to 24 VDC.

Breakthroughs in technology have allowed the size of LNBs to decrease. As a result, manufacturers have had the opportunity to combine higher performance with lower cost. Even with the smaller sizes, LNB designs continue to have specific characteristics such as the shape of the waveguide. An LNB is not considered as a repairable item; manufacturers hermetically seal the electronic components to prevent damage from moisture.

Low Noise Amplifier Temperatures

As a result, the low noise amplifiers used for satellite systems are measured according to noise temperature in degrees Kelvin. A low noise temperature signifies that the amplifier circuitry will introduce less noise into the signal. Today, typical C-band amplifiers have noise temperature ratings that range from 17°K to 25°K. During the 1980's, noise temperature ratings for low noise amplifiers ranged from 120°K to 150°K.

Noise Factors, Temperature, and LNB Gain

Because the LNB works as the first amplifier, or head-end, in the satellite system, it can add noise to the received signal. In chapter three, we calculated the total system noise temperature through the use of the noise temperature of the LNB. Modern C-Band low-noise block converters can have noise figures as low as 0.35 dB. The noise temperature measures the amount of noise in degrees Kelvin. Using temperature to measure the amount of noise may make more sense when we consider that internal noise within a semiconductor occurs because of molecular movement. If we decrease the operating temperature of a semiconductor to absolute zero degrees Kelvin, molecular movement stops. As a result, an LNB with a lower noise temperature has better noise characteristics than an LNB with a higher noise temperature.

In contrast, the noise figure measures the ability to maintain a usable satellite signal as the signal passes through the LNB. We can convert the noise figure to the noise temperature with the following equation:

$$\text{Noise Temp or T in degrees Kelvin} = 290\ (10^{[10\log10\ /\ 10]} - 1)$$

As the temperature rises, the noise figure also rises. *Figure 5.17* shows how the noise figure varies with increases in ambient temperature.

As mentioned earlier, manufacturers grade low noise amplifiers used for satellite systems according to noise temperature and degrees Kelvin. Every 20° drop in amplifier noise temperature produces an approximate increase of 0.6 decibels of gain. The typical gain given by an amplifier with a noise temperature of 20° to 25° Kelvin is 65 decibels. Modern low noise amplifiers rely on gallium arsenide field affect transistors, or GaAsFETs, to hold the increase the gain while reducing noise to the lowest possible level. GaAsFET amplifiers cut the internal noise within the LNB to a minimum because of characteristics that allow the amplifier to act as if it has an operating temperature of near absolute zero.

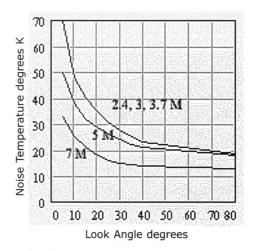

Figure 5.17. Illustration of noise characteristics vs. temperature increases.

Most LNBs have gain figures in the 50 to 70 dB range. The high signal gain occurs because each amplifier stage within the device adds 10 to 12 dB of gain. With such high gain stages, the level of noise given by the later stages will remain negligible. The high gain figures occur without overloading or intermodulation because the satellite signals have similar levels. In addition, the high level of gain overcomes cable losses and the noise contributed by each amplifier stage and the block downconversion stages.

Because of these factors, the gain of an LNB must always exceed a minimum value. As a rule, 50 dB of gain stands as the minimum amount needed to overcome the typical losses seen found within a system. The amount of gain given by an LNB changes with both temperature and signal frequency. Every 10° centigrade increase in temperature can decrease the amount of gain by 0.6 dB. Given the selectivity of the amplifiers within the LNB, high gain occurs only for desired frequencies; the gain decreases for frequencies found outside the desired frequency band.

Attaching the LNB to the Feedhorn

In installations that utilize a button hook feed support, we can save time by assembling the entire feed, LNB, and support before installing everything on the dish. Looking at *Figure 5.18*, a rectangular flange on the back of the feedhorn fastens precisely with a similar flange found on the LNB. All portions of the rectangular waveguide of the feedhorn assembly must match with the waveguide assembly of the LNB. When attaching the feedhorn to the support bracket, ensure that the wires exiting from the polarizer portion of the feedhorn point downward.

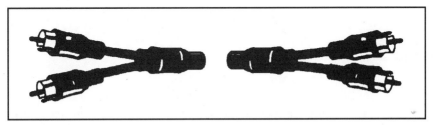

Figure 5.18. RCA jack connectors.

When attaching the LNB to the feed, always insert the rubber weather seal between the LNB and feed mounting plates and ensure that the seal remains in place. The weather seal prevents moisture from entering the feedhorn, waveguide, and LNB, and it keeps moisture from degrading the signal reception or damaging components. Most manufacturers supply the bolts, washers, and nuts used to fasten the LNB and feedhorn together. *Figures 5.19A*, *5.19B*, and *5.19C* illustrate the steps for properly installing the feed, servo motor, and LNB.

Figure 5.19A. Installing the waterproof seal at the waveguide.

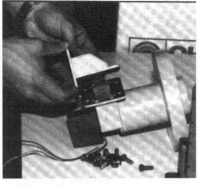

Figure 5.19B. Attaching the 90 degree waveguide to the feed.

Figure 5.19C. Attaching the LNB to
the feed assembly.

Cables

Because a satellite system usually includes a servo motor for moving the polarizer probe, an actuator for moving the reflector dish, and electronics for amplifying and manipulating signals, a number of different cables connect between the devices. Rather than purchase each cable separately, most installers use all-in-one ribbon cables that include two transmission cables, a set of electrical cables designated for the polarizer, larger gauge electrical cables designated for the actuator motor, and smaller gauge electrical cables used for the actuator controller. Most all-in-one ribbon cables have a polyurethane jacket that allows the direct burial of the cables without the use of any type of conduit.

Because water and animals can quickly damage transmission and electrical cables, many installers insist on enclosing the ribbon cable within either PVC plastic pipe or aluminum conduit. Every LNA, LNB, LNC, or LNF will have an IF connector at the back of the device. After attaching the coaxial cable to the connector, always wrap the connection with a waterfproofing compound. The sealant will prevent moisture from entering both the LNB and cable connection.

Transmission Cables

Without the use of proper cabling, the high frequency microwave signals found in a satellite receiving system would radiate away from the system. In addition, desired signals would become vulnerable to degradation caused by external noise interference. Signals from the low noise amplifier or low noise block downconverter travel to the receiver and then to a television or other devices through coaxial cables. Shown in *Figure 5.20*, a coaxial cable consists of a copper center conductor insulated from an aluminum or copper outer conductor by a dielectic material and then covered by a weatherproof jacket. The center conductor carries the signal while the outer conductor both prevents the transmitted signal from interfering with other portions of the system or other electronic equipment and eliminates most interference or noise from external sources.

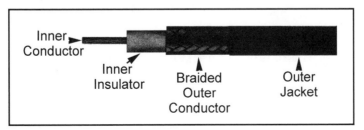

Figure 5.20. Coaxial cable.

Coaxial cable options include the hardline coax which has a rigid metal sheath, conventional coax which utilizes a braided copper or aluminum sheath, and foam coax which uses either foam or compressed air as the dielectric material. Hardline and conventional coaxial cables feature either a polypropylene or polyethylene dielectric. In terms of transmission and shielding, hardline coax cables transmit high frequency signals and protects the signal from loss through the use of the metal sheath. The use of foam or compressed air as a dielectric in foam coax cables proves even lower loss figures.

Chart 5.2 provides additional information about different types of coaxial cables and compares the signal carrying capacities of coax cable types and sizes. Every type of cable used for the transmission of signals has characteristics that can aid or impede the flow of the electrical current that makes up the signal. Those factors include characteristic impedance, carrying capacity in terms of distance, and the composition of the conductors and shields.

Connectors

The use of different cables within the satellite system also means that the system will require the use of different types of cable connectors. While the coaxial cables needed for the transmission of the signals from the amplifier to the receiver require F-connectors, other types of connectors work for the polarizer and actuator cables. Each type of cable connector also requires the proper installation to limit attenuation, power losses, and losses due to characteristic impedance.

Coaxial Cable Type and Size	Loss in Decibels per 100 Feet	Signal Carrying Capacity (frequency and distance)	Application
RG-59	-3 dB at 70 MHz	0-950 MHz at 100 Feet	Residential Cable Installation 75 Ohms
RG-6	- 2.2 dB at 70 MHz	0-1750 MHz at 100 Feet	Residential Cable Installation 75 Ohms; Block Downconversion 75 Ohms
RG-11	-1.9 dB at 70 MHz	0-1750 MHz at 200 Feet	Block Downconversion 75 Ohms
RG-213	-22 dB at 4 GHz	0-16 GHz at 10 Feet	Microwave Transmission 50 Ohms
RG-214	-20 dB at 4 GHz	0-16 GHz at 50 Feet	Microwave Transmission 50 Ohms
RG-217	-17 dB at 4 GHz	0-16 GHz at 50 Feet	Microwave Transmission 50 Ohms
1/2 Inch Hardline	-8 dB at 4 GHz	0-16 GHz at 100 Feet	Main Trunk Line of Cable TV Network 50 Ohms
7/8 Inch Hardline	-6 dB at 4 GHz	0-16 GHz at 100 or More Feet	Main Trunk Line of Cable TV Network 50 Ohms

Table 5.2. Coaxial cable types and performance.

F-Connectors

Pictured in *Figure 5.21A*, F-connectors fasten onto coaxial cable leads. The female connection found at the cable connects to a male F-connector found at the signal input of a television receiver, stereo receiver, satellite receiver, or LNB. Different sized F-connectors work for either RG-59 or RG-6 coaxial cable and install through the use of a crimping tool.

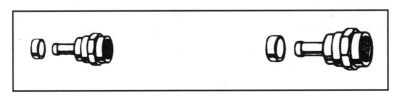

Figure 5.21A. F-connectors.

Referring to *Figure 5.21B*, install an F-connector by stripping approximately 3/4" of the outer insulation from the coax. Next, cut the shield lead to a length of approximately 3/8" and fold the shield back over the outside insulation. Remove approximately 3/8" of the insulation from the inner conductor and gently scrape the conductor to ensure good electrical conductivity. Slide the F-connector onto the cable so that the inner insulation remains flush with the inside shoulder of the connector. Then, use the crimping tool to securely fasten the connector to the shield wire and outer insulation.

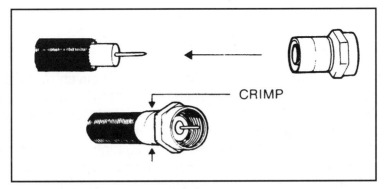

Figure 5.21B. Installing an F-connector onto a coaxial cable.

Other Connectors

Many times, RCA jacks such as those pictured in *Figure 5.22* connect the audio and video output signals from a satellite receiver to a VCR or television monitor. Because of the lack of shielding found with RCA connectors, the connectors will not work for high frequency applications. RCA connectors solder onto the transmission cable. Scotch locks and solderless lugs provide an inexpensive, easy method for connecting two wires together. Scotch locks work well with polarizer cables while solderless lugs attach wires to screw-on terminals. BNC connectors and DIN connectors provide another option for connecting cables to equipment. BNC connectors are often used with VCRs for the video IN and video OUT connections. DIN connectors attach multiple wires to audio/video equipment.

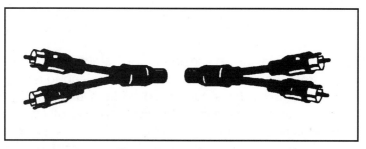

Figure 5.22. RCA jacks.

Cable Signal Loss

Attenuation

As we saw in chapter two, attenuation is the opposite of gain and is shown as loss in terms of decibels. With attenuation, the output signal from an electronic circuit has a lower amplitude or signal level than the input signal. Attenuation becomes more of a factor when high frequencies are transmitted throughout a system. The lack of attenuation exists as one of the key benefits given through the combining of the low noise amplifier and block downconverter within a single LNB package. In addition, minimizing the length of cable runs and limiting the angles of any cable bending also lessens attenuation.

Line Amplifiers and Line Splitters

In some instances, the length of coaxial cable between the feed and the receiver will exceed the signal carrying capacity of the cable and allow signal loss to occur. A *line amplifier* builds additional gain into the system and compensates for the loss. Shown in *Figure 5.23*, a line amplifier uses the small dc voltage traveling along the center conductor of the coaxial cable as a power supply and usually includes a microwave filter.

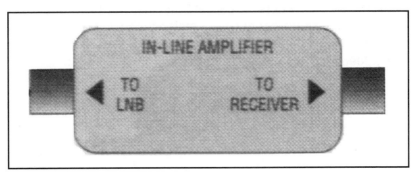

Figure 5.23. A line amplifier.

A *line splitter* divides the signal along two or more paths. Pictured in *Figure 5.24*, line splitters are designated according to frequency ranges. A splitter found between the LNB and receiver handles frequencies between 950 and 1450 MHz. Line splitters used between the receiver and television divide lower range frequencies.

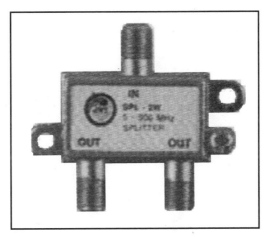

Figure 5.24. A line splitter.

Summary

Chapter five addresses the installation of the feedhorn, polarotor, and low noise amplifier on the satellite system. As the chapter moves through each stage of the installation, it combines theory with practice as it uses equations to explain operation. The emphasis on practical skills continues with instructions about installing the feed, polarotor, and amplifier.

During the description of low noise amplifiers, the chapter takes the reader through the evolution of low noise amplifier and conversion technologies. The tour concludes with an explanation of low noise block downconversion and a description of modern low noise block feeds that combine the feedhorn, polarity switching, and amplification within one device. Chapter five also provides information about cables used for the polarotor and LNB as well as connectors.

CHAPTER

An Inside Look at Satellite Receivers

Modern satellite communications systems relay a larger selection of television programming than ever seen. As you select a satellite receiver, the choice will depend on the programs that you wish to view. While you have a broad range of "free-to-air" programs to select from, other programs are available only through paid subscriptions. The subscription-only programming includes the major networks, superstations, movie channels, and sports broadcasts.

In addition to the range of viewing choices, many broadcast services transmit only in an analog format while others transmit programming only in a digitally compressed format. As you know, analog signals consist of a continuous wave of information while digital signals consist of bytes of information that represent "on" and "off" conditions for logic circuits.

Any type of satellite television receiver provides a specific set of functions. Those are:

- providing voltages for the LNB, the actuator, and the remainder of the receiver

- accepting modulated RF signals

- selecting a station

- converting the signal allocated for that station to an intermediate frequency

- amplifying and filtering the IF signal

- detecting or demodulating the signal

- amplifying the final signal after no additional frequency conversions occur

- supplying the signal for use by the customer

Satellite television receivers include all these functions but differ from typical receivers—such as those used for the reception of AM/FM radio signals—in the distribution of the functions within the unit. As an example, part of the amplification and conversion of the incoming signals occurs at the low-noise block downconverter.

Because of the programming choices and different market applications, TVRO receivers also offer a wide range of features and options that can match any consumer desire. While some consumers may remain satisfied with a "no frills" approach, others may elect to spend additional money for receivers that provide the capability to receive C and Ku-band transmissions along with a combination of analog and digital transmissions. Most modern receivers combine the receiver, decoder, and the antenna positioner into one unit and include some type of remote control.

Voltage Supplies for the LNB, the Actuator, and the Remainder of the Receiver

In addition to processing and converting information, satellite receivers also provide voltage and current that allow the consumer to select from a wide range of available channels. Current flows through the coaxial cable attached between the feed and the receiver and powers the low noise amplifier and downconverter. The receiver also provides the voltages needed for polarity switching through the ground, voltage, and pulse connections. Other cable connections attach the controller portion of the receiver to the actuator and provide the voltages need to both move the actuator and control its position.

The conversion of signals within the satellite receiver can only occur through the application of dc voltages throughout the receiver. As indicated in chapter one, each of the dc voltages works as a source voltage for a particular circuit. Power supplies convert the pulsating ac voltages found at electrical outlets into the regulated dc voltages required by the individual circuits. The power supply used in a satellite receiver also provides a range of voltages needed for tuning, the audio and video processing circuits, the tuner/demodulator module, digital circuits, displays, and the polarizer control.

Voltage Supplies

Every stage in an electronic device requires some type of voltage supply because of the signal amplification required to make the system function. Amplification is an increase in the voltage, current, or power gain of an output signal. Although systems may utilize an ac power line input, the components within the system rely on dc voltages. We can categorize dc voltages into the low (12-35VDC), medium (150-400VDC), and high (15-25kVDC) ranges. Using a television receiver as an example, the low voltage supply provides the necessary voltages for semiconductor operation while the mid-range and high voltages are required for the deflection, focus, and CRT circuits.

Every electronic voltage supply has four distinct parts that involve rectification, regulation, and filtering of the source voltages. Transformers either step up or step down the voltages to the levels required by circuits. Rectification involves the conversion of the required ac voltage value to a pulsating dc voltage. Regulation is defined as the maintenance of a consistent output at the power supply source. With a regulated power supply, changes in the input voltage do not affect the operation of some stages in the system. Filtering smoothes the pulsating dc voltage into a usable, constant dc supply voltage.

Linear Power Supplies

Depending on the manufacturer, a satellite television receiver may have either a linear power supply or a switched-mode power supply. Linear power supply functions conform to the basic block diagram shown in *Figure 6.1* and have the basic purpose of converting an ac line voltage to dc voltages. Most linear power supplies feature a power transformer that steps down the 115/230VAC 50/60Hz line voltage to the lower voltages required by semiconductors and isolates the load from the ac line input. However, depending on the application, a power supply also may operate as a transformerless system.

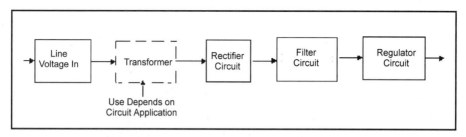

Figure 6.1. Block diagram of a linear power supply.

Transformerless power supplies are found in ac-dc receivers and in nearly all video equipment. The lack of a transformer in the power supply cuts the cost for manufacturing the unit but eliminates the isolation between the ac line input and the load. As a result, transformerless systems require a grounding system that does not rely on a chassis ground.

Regardless of whether the application utilizes a power transformer or not, rectifiers convert the ac line voltage to pulsating dc voltages while filters smooth the ripples found in the rectified dc voltage. The filter circuits may consist of either capacitors, inductors, resistors, or a combination of passive components. Linear power supplies also feature a regulator stage which is designed to maintain the power supply voltage at constant levels despite any load current variations.

In brief, linear power supplies provide excellent regulation, simple circuit designs, a relatively fast transient recovery time, and low output ripple. However, the drawbacks of linear power supplies include a lower-than-desired 50% efficiency rating because of the amount of power wasted while converting an unregulated dc voltage into a lower, regulated dc voltage. Much of the power loss occurs during high line voltage/high load conditions.

Linear Power Supplies that use Transformers

Two fixed coils wound on a core make up a transformer that transfers electric energy from one or more circuits by electromagnetic induction. The inducing of a voltage in the transformer secondary depends on the building and collapsing of a magnetic field. Because of this, the primary windings of a transformer always connect to an intermittent source of current or to a source of alternating current while the secondary connects to the load.

Linear power supplies using a power transformer provide *isolation* between the ac line voltage and the load. In this case, a thin-layer of insulation separates the primary and secondary windings. Moreover, either side of the secondary windings may connect to the chassis ground while one side of the primary windings connects to earth ground.

Transformerless Linear Power Supplies

Most electronic equipment and all electronic devices using ac/dc power supplies do not utilize a power transformer for the stepping up of the voltage. While those supplies provide some benefits such as cost savings, the disadvantage lies within the inability to provide a voltage that has an amplitude higher than the peak of the ac line voltage. A transformerless design has no isolation between the chassis and ground and relies on a grounding system called the hot chassis.

Rectifying the AC Line Voltage

A diode is a two-terminal semiconductor device that conducts under specific operating conditions. PN junction diodes are constructed of an n-type material at the cathode and of a p-type material at the anode. The n-type material is more negative than the p-type material. An ideal diode acts as an open switch when reverse biased and as a closed switch when forward biased.

Rectification involves the conversion of the required ac voltage value to a pulsating dc voltage that may have either a positive or negative polarity. Depending on the application, linear power supplies may use one of four ac-to-dc voltage diode rectifier circuits to deliver either a half-wave or a full-wave output. Those rectifier circuits are:

- the half-wave rectifier circuit

- the full-wave rectifier circuit

- the full-wave bridge rectifier circuit

- the direct-coupled doubler circuit

Half-Wave Rectifiers

Used with hot chassis ground systems in small-screen television receivers, the half-wave rectifier provides the simplest method for rectifying an ac voltage. As shown in *Figure 6.2A*, the transformer supplies the desired ac voltage to the diode. Connected in series with the T1 secondary and the RL, representing the circuit load, the diode conducts when its anode is positive with respect to the cathode.

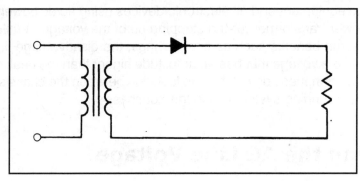

Figure 6.2A. Schematic diagram of a half-wave rectifier.

Figure 6.2B links the conduction of the diode with the alternations of the ac sine wave. During a positive alternation of the ac wave, the top of the T1 secondary is positive with respect to the bottom windings. As a result, the diode conducts and current flows through the load. During the opposite alternation, the top of the secondary is negative with respect to the bottom windings and the diode does not conduct. Because the diode only conducts on positive alternations, the current pulses flow through the load in only one direction. Thus, the circuit has a pulsating dc output voltage.

Figure 6.2B. Waveform of the half-wave
rectifier output.

Full-Wave Rectifiers

Half-wave rectifiers offer the benefits of simplicity and low cost. However, most modern circuit designs cannot rely on the pulsating dc output voltage produced during each positive alternation of the ac wave. Instead, many designs utilize a full-wave rectifier in the power supply. The full-wave rectifier always features a center-tapped transformer, develops two current pulses, and is used in cold chassis designs.

Referring to *Figure 6.2C*, the center-tapping of the transformer establishes two half-wave circuits. The diode connected to the top half of the T1 secondary conducts only when the top-half is positive with respect to the center-tap while the diode connected to the bottom half of the secondary conducts only when the bottom-half of the secondary is positive with respect to the center-tap. Each alternation of the ac sine wave shown in *Figure 6.2D* forward biases and then reverse biases the diodes. As a result, the full-wave rectifier circuit produces two current pulses that flow through the load on each positive and negative alternation of the ac sine wave.

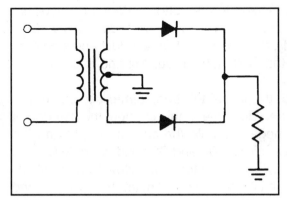

Figure 6.2C. Schematic of a full-wave rectifier.

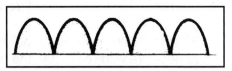

Figure 6.2D. Waveform of a full-wave
rectifier output.

Full-Wave Bridge Rectifiers

Pictured in *Figure 6.2E*, full-wave bridge rectifiers use power transformers that do not have a center-tapped secondary to provide full-wave rectification. Instead, the full-wave bridge features four diodes that rectify the full secondary voltage on each alternation of the ac wave. Compared to the full-wave rectifier diagrammed in *Figure 6.3A*, the bridge rectifier provides pulsating dc with almost twice the amplitude while using a secondary with the same amount of turns. The pulsating dc output almost equals the peak voltage found across the transformer secondary. Full-wave bridge rectifiers are used in cold and hot grounding systems and are generally applied to large-screen televisions.

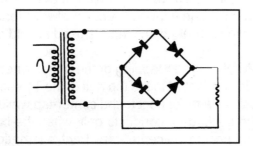

Figure 6.2E. Schematic of a full-wave bridge rectifier.

While referring to *Figures 6.2D* and *6.2E*, you can begin your analysis of the bridge rectifier operation with the first positive alternation of the ac wave making the top of the transformer secondary positive with respect to the bottom. With this condition, the cathode of D1 and the anode of D4 are positive and the cathode of D3 and the anode of D2 are negative. Consequently, diodes D3 and D4 conduct and electrons flow from point "B" through D3 through the load then through D4 and to point "A."

On the other alternation of the ac wave, the top of the transformer secondary becomes negative with respect to the bottom. Because of this change, the cathode of D1 and the anode of D4 also become negative while the cathode of D3 and the anode of D2 become more positive. Thus, diodes D1 and D2 conduct while D3 and D4 do not. The conduction of D1 and D2 allows electrons to flow from point "A" through D1, through the load, then through D2, and finally to point "B." With D1 and D2 conducting on one alternation and D3 and D4 conducting on the other, the circuit produces two current pulses for each cycle of the ac voltage supply and provides full-wave rectification.

Integrated Rectifiers

Advances in the production of integrated circuits have allowed manufacturers to place all the active components of a full-wave or bridge rectifier into one semiconductor package. An example would be that a single one inch square package can contain a bridge rectifier capable of handling an average forward current of 25mA and surges as high as 400 amps. The performance of an integrated rectifier package is equivalent or superior to that provided by conventional diode rectifier circuits.

Filtering

Every type of rectifier circuit uses some type of filter circuit to smooth the pulsating dc output voltage given through rectification and to remove as much rectifier output variation as possible. The most basic type of filter consists of an electrolytic capacitor connected across the output of a half-wave rectifier. Other filter types take advantage of the reactance properties of inductors or utilize a combination of a power transistor and a capacitor.

Capacitor Filters

Referring to *Figures 6.3*, when the first rectifier current pulse begins to increase, electrons flow into the lower plate of C1, follow the direction shown by the charge arrow, flow out of the upper capacitor plate, flow through diode D1, and through the secondary of the transformer. As the transformer secondary voltage decreases, the capacitor discharges. However, because D1 allows electrons to flow in only one direction, the capacitor cannot discharge through D1 and the transformer secondary. Instead, the discharge current flows from the lower plate of the capacitor through the load and to the upper capacitor plate. Because the capacitor value does not allow a rapid discharge, the load current and capacitor voltages do not decrease as quickly as the secondary voltage.

Once the secondary voltage decreases to a level less than the voltage across C1, D1 conducts only long enough to charge the capacitor before becoming reverse biased. Thus, C1 can only partially discharge before another ac pulse begins the recharging cycle and causes the capacitor to charge to the peak of the secondary voltage. All this smoothes any pulses found across the load. We define the varying voltage across the filter capacitor as the *ripple voltage.*

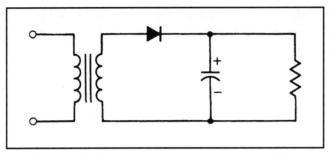

Figure 6.3. Schematic diagram of a capacitor filter.

Inductive Filters

Inductive filters protect the power supply circuit from current surges. Improvement of the filtering action occurs through the addition of a filter choke, an inductor consisting of a winding on an iron core, and another capacitor. The self-induction of the choke opposes any changes in current and when connected in series with the rectifier and load, it carries the entire load current.

LC Filters

Figure 6.4 shows the addition of the choke to the half-wave rectifier circuit. When the applied secondary voltage causes an increase in current, the self-induced voltage within the choke opposes the change and prevents the current from increasing immediately to its peak. As the filter capacitor across the input of the circuit charges to the peak of the applied voltage during the first current pulse and the voltage across the transformer secondary begins to decrease, the input capacitor discharges through the load and the coil. The decreasing load current induces a voltage in the choke in a direction that maintains the current amplitude.

Figures 6.5A, 6.5B, and *6.5C* show how the waveforms across the capacitors and inductor should appear. While the first figure shows the waveform found across the input capacitor, the second shows the waveform at the inductor. Because the choke has a very low resistance, almost no voltage drops across the coil and the counter emf of the inductor opposes any ac ripple. *Figure 6.5C* shows the smooth dc output voltage found across capacitor C2.

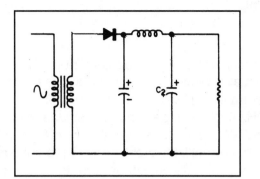

Figure 6.4. LC filter circuit.

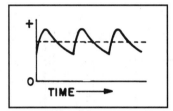

Figure 6.5A. Waveform across the input capacitor.

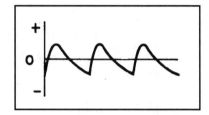

Figure 6.5B. Waveform across the inductor.

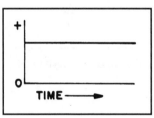

Figure 6.5C. Waveform across the second capacitor.

RC Filters

Some applications that require only a small current through the load replace the filter choke with the filter resistor shown in *Figure 6.6*. R1, the filter resistor, connects in series between the rectifier and the load and forms the series arm of a filter consisting of C1, R1, and C2. In the figure, the voltage EC1 is applied to R1 and C2 in series. Because capacitor C2 passes ac frequencies and blocks dc voltages, part of the direct voltage and a larger part of the ac line voltage appear across the filter resistor.

Active Filters

Many power supply circuit designs will supplement a filter capacitor with an active power filter circuit. Usually consisting of either a single power transistor or a combination of a filter driver transistor and a power transistor, the active filter circuit eliminates 60 Hz and 120 Hz ripple voltages and any residual audio frequency or horizontal frequency voltages. With the active filter shown in *Figure 6.7*, a power transistor connects in series with the rectifier output. Any ripple voltages cause increases in the current flowing through the transistor. In turn, the increased current causes an increase in the reverse bias of the transistor and decreases the amount of conductance. As a result, the voltage at the output of the transistor also decreases.

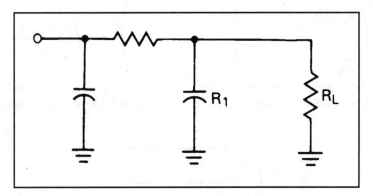

Figure 6.7. Active filter.

Regulation

Regulation is the maximum change in a regulator output voltage that can occur when the input voltage and load current vary over rated ranges. We can also define regulation in terms of line regulation and load regulation. While *line regulation* of a voltage regulator shows the amount of change in output voltage that can occur per unit change in input voltage, the *load regulation* rating of a regulator indicates the amount of change in output voltage that can occur per unit change in load current. An ideal voltage regulator will maintain a constant dc output voltage despite any changes that occur in either the input voltage or the load current.

Regulation is required because every device in the circuits that feeds off the low voltage supply has an internal resistance and draws some amount of current. Without some type of voltage regulation, the combination of internal resistance and current flow creates a voltage drop across the resistance and a resulting decrease in the output voltage. Regulator circuits stabilize the rectified and filtered power supply voltages so that the dc level of the voltage does not vary with changes in the line or load.

The basic types of regulators are the shunt regulator and the series regulator. A *shunt regulator* is in parallel with the load while a series regulator is in series with the load. Zener diode regulators are a form of shunt regulators. Transistors are a form of series regulator. Both types of particular applications have advantages and disadvantages.

Regulator circuits for linear power supplies range from a zener diode, transistor circuits working as series-pass regulators, silicon-controlled rectifier (SCR) circuits working as phase-control regulators, and integrated three-terminal regulators. Most satellite receivers utilize hybrid regulators for voltages in the +5.1VDC to +12VDC range.

Zener Diode Regulators

Referring back to chapter one, a zener diode is a special type of diode that operates in the breakdown region of the diode characteristic curve and often works as a voltage regulator. The most basic type of voltage regulation occurs through the use of a zener diode and takes advantage of the reverse breakdown characteristic of the diode. That is, when a zener diode operates in the reverse breakdown region, the diode will have a constant voltage across it as long as the zener current remains between the knee current and the maximum current rating.

As shown in *Figure 6.8*, a zener diode regulator circuit places the load resistance in parallel with the diode. Therefore, the load voltage will remain constant as long as the zener voltage remains constant. If the zener current increases or decreases from the allowable range, the zener and load voltages change. Consequently, the success of a zener diode regulator in maintaining the load voltage depends on keeping the zener current within its specified range.

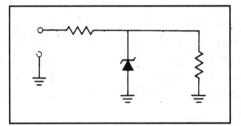

Figure 6.8. Zener diode regulator.

Integrated Circuit Three-Terminal Regulators

Many modern power supplies rely on an IC voltage regulator because of the characteristics provided by the IC technology. Three-terminal IC voltage regulators hold the output voltage from a dc power supply constant over a wide range of line and load variations. IC 3-terminal regulators include the:

•   Fixed-positive voltage regulator

•   Fixed-negative voltage regulator

•   Adjustable voltage regulator

•   Dual tracking regulator

The first two types of IC regulators provide specific positive or negative output voltages. An adjustable IC regulator allows either a positive or negative output voltage to adjust within specific limits. A dual-tracking IC regulator establishes equal positive and negative output voltages. Depending on the application, different types of dual tracking regulators may provide fixed-output voltages or output voltages that adjust between specific limits.

While IC voltage regulators are used in many power supply circuits, several limitations exist. When using the fixed and adjustable voltage regulators, the polarity of the input voltage must match the polarity of the output voltage. With the dual-tracking regulator, the IC must have dual input polarities. Also, any IC voltage regulator must include a shunt capacitor for the prevention of oscillations and an output shunt capacitor to improve ripple rejection.

Voltage Dividers

A *voltage divider* establishes a method for providing more than one dc output voltage from the same power supply. The most basic voltage divider places a number of resistors across the power supply terminals. For example, the three-resistor voltage divider can provide three different output voltages. The values of the resistors are chosen to accommodate the amount of current required by the load connected to the particular terminal.

Switched-Mode Power Supplies

All modern televisions, monitors, personal computers, VCRs, and many other types of electronic equipment rely on a different type of power supply called the switch-mode power supply, or SMPS. *Switched-mode power supplies* offer advantages such as reduced size, weight, and cost. The high frequency operation of a SMPS allows the use of smaller and lighter components than those seen in linear power supplies. In addition to those benefits, an SMPS offers greater efficiency than a linear power supply. Because an SMPS operates either fully on or fully off, this type of supply loses little power and has an efficiency of approximately 85%.

SMPS Basics

When we considered linear power supplies, we looked at a block diagram that showed a line input voltage traveling into a power transformer and then through a rectifier circuit, filter, and regulator circuit. With switch-mode power supplies, the block diagram changes slightly. Rather than begin with a transformer, the SMPS begins with a full-wave rectifier circuit connected directly to the line and then progresses to a high-frequency transformer, a power transistor, and a pulse generator. While *Figure 6.9* shows a block diagram for a typical switch-mode power supply, *Figure 6.10* features a schematic diagram of the SMPS. The SMPS supplies 132VDC for the sweep circuits, 12VDC for a remote control preamplifier, 12VDC for the turn-on, and 35VDC for the audio stages of a television receiver.

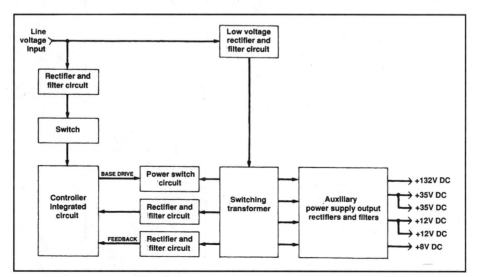

Figure 6.9. Block diagram of an SMPS.

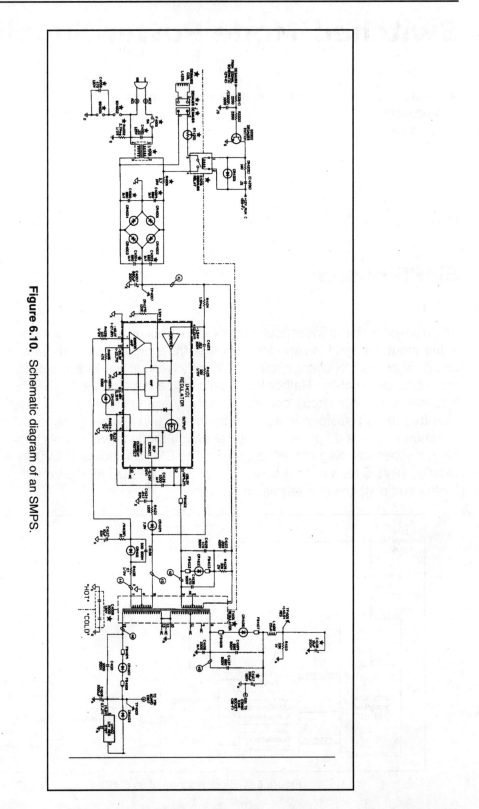

Figure 6.10. Schematic diagram of an SMPS.

SMPS Components

As with the linear power supplies studied in chapter two, switched-mode power supplies contain a mix of passive and active components. Those include bipolar junction transistors, rectifiers, silicon-control rectifiers, shunt regulator ICs, optoisolators, filter and bypass capacitors, resistors, metal oxide resistors, and thermistors. Each individual component type affects the performance of the switching power supply and involves tasks such as feedback, control, rectification, overvoltage and overcurrent protection, regulation, isolation, filtering, and voltage division.

When studying the operation of an SMPS, it is easier to consider the components by function. For example, filter capacitors either filter the rectified—and sometimes doubled—ac line input voltage or filter the output voltages from the SMPS. Other types of capacitors in the circuit provide bypass paths. SMPS power supplies also contain a combination of general type resistors and flameproof resistors, metal-oxide varistors, and thermistors. While the general type resistors are often found in voltage divider circuits, the flameproof resistors are found in the return circuit for the switching regulator or in the ac line circuit. Metal-oxide varistors and thermistors provide protection against severe surges and appear in the ac line circuits while opto-isolators or opto-couplers establish isolation.

Active components such as bipolar transistors, MOSFETs, and SCRs may operate as part of a feedback circuit, as regulators, or in overvoltage and overcurrent protection circuits. Bipolar junction transistors work as either components in a feedback circuit or may function as the SMPS switching device. The type of transistor used in the particular circuit varies with the function. For example, a power transistor capable of handling high voltages will work as a switching device.

In addition to transistors, MOSFETs and SCRs may appear in the switching role. SCRs are also found in overvoltage and overcurrent protection circuits. Rectification occurs through the use of either discrete or packaged diodes. Most SMPS units use diodes for ac line rectification or in voltage doubler circuits. The switched power supplies usually rely on some type of 3-pin IC regulator for regulation of the output voltages.

SMPS Operation

All switched-mode power supplies use a high frequency switching device such as a transistor, MOSFET, or SCR to convert the directly rectified line voltage into a pulsed waveform. An SMPS that has a lower power requirement will feature a conventional transistor or MOSFET as a switcher while high power SMPS units will rely on a SCR, or triac. Each of the last components offers latching in the on state and high power capability. However, this type of capability also requires more complex circuitry to ensure that the semiconductors turn off at the correct time.

The switching on and off of the transistor closes and opens a path for dc current to flow into the transformer. With the flow of current producing a changing magnetic field in the transformer primary, a changing magnetic field also develops in the transformer secondary winding. As a result, voltage is induced in the secondary winding. Rectifiers and filters in the secondary circuit rectify and filter into stable supply voltages.

SMPS Input

After the rectification of the line voltage, the SMPS may have two possible dc inputs. With the first, 150-160VDC arrives at the SMPS after the direct rectification of 115-130VAC line voltage. However, some SMPS units require a higher input voltage. In this case, a voltage doubler supplies 300-320VDC to the SMPS input. Other designs rectifiy a 220-240VAC line voltage and also supply the 300-320VDC to the SMPS input.

While rectification of the line voltage occurs through the use of a full-wave bridge rectifier or a voltage doubler, the input to the SMPS also includes inductors and capacitors for the purpose of filtering line noise and any voltage spikes. Those components also eliminate the transmission of any radio frequency interference generated by the power supply back into the ac line. As mentioned, most designs feature metal-oxide varistors across the input lines for additional protection against surges.

Switched-Mode Regulators

Switched-mode regulator circuits provide the advantage of having a control device that has minimal power dissipation for the entire duty cycle. In particular, switched-mode regulator circuits provide:

- The capability to produce an output voltage higher than the input voltage

- The capability to produce either a positive or negative output voltage from a positive input voltage, and

- The capability to produce an output voltage from a dc input voltage

A switched-mode regulator circuit uses a control device—such as a bipolar transistor, a field-effect transistor, or a silicon-controlled rectifier—to switch the supply power in and out of the circuit and regulate the voltage. Switching occurs because of the ability to send the device into either saturation, the completely-on state, or into cutoff, the completely-off state.

The duty cycle of the device, or the ratio of "on" time to "off" time, establishes the regulation of the output voltage level. Therefore, regulation in a switched-mode power supply occurs through the pulse-width modulation or the pulse-rate modulation of the dc voltage. *Pulse-width modulation* varies the duty cycle of the dc voltage while *pulse-rate modulation* varies the frequency of the dc pulses.

Figure 6.11 provides an illustration of pulse-width modulation. In the figure, the on-cycles of the pulse train energy double as the time periods for storing energy in a magnetic field. During the off cycles of the pulse train, the stored energy provides output power and compensates for any changes in the line voltage or the load. The pulse-width modulation of the switching transistor changes the conduction time of the device by varying the pulse frequency.

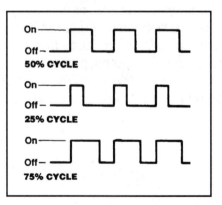

Figure 6.11. Illustration of pulse-width modulation.

With a low load resistance and the line voltage within tolerance, the switched-mode power supply switches the power into the power supply for only a short period of time. Either a high load resistance, a low line voltage, or a combination of both conditions will cause the switched-mode power supply to transfer more energy over a longer period of time into the power supply. As a result, the switching frequency varies from a higher frequency for lower loads to a lower frequency for higher loads.

SMPS Transformer Operation

Switched-mode power supplies do not include any type of conventional power transformer and, as a result, do not have line isolation. At the input of the power supply, a small, high frequency transformer converts the pulsed waveform taken from the switching device into one or more output voltages. Other components following the high frequency rectify and filter the voltages for use by signal circuits.

Isolation in the SMPS System

Although the SMPS does not provide line isolation, the use of the high frequency transformer establishes an isolation barrier and the type of characteristics needed to operate in the flyback mode. Depending on the circuit configuration, a small pulse transformer or an opto-isolator sets up feedback across the isolation barrier. The feedback controls the pulse width of the switching device and maintains regulation for the primary output of the SMPS.

Most small switched-mode power supplies such as those used for satellite receivers use opto-isolators for feedback. An *opto-isolator* is a combination of an LED and a photodiode in one package and establishes an isolation barrier between low voltage secondary outputs and the ac line. Whenever a primary output voltage reaches a specified value, a reference circuit in the output turns on the LED. In turn, the photodiode detects the light from the LED and reduces the pulse width of the switching Y. This establishes the correct amount of output power and maintains a constant output voltage. Along with the primary output winding, the transformer has 6 or more separate windings that provide positive and negative voltages for the electronic system.

Ground

The Cold Chassis Ground

A *cold chassis ground* has any point on the metal chassis at ground or "earth" potential. Receivers built around the cold chassis rely on either a voltage regulating transformer or an active device as a power transformer. With the cold chassis, the ac voltage line connects directly to the power transformer primary. Then, the secondary of the transformer provides ac waveforms for the power supplies located throughout the receiver. In this type of chassis, the space between the primary and secondary windings of the transformer serves as the sole point of power line isolation for the receiver. *Figure 6.12* shows a schematic of a typical cold chassis ground system.

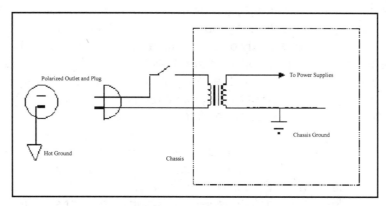

Figure 6.12. Cold chassis ground.

The Hot Chassis Ground

As *Figures 6.13* and *6.14* illustrate, different symbols represent a cold and hot ground. The traditional ground symbol seen in the *Figure 6.13* represents the earth ground while the inverted open triangle shown in *Figure 6.14* represents the hot ground. With the hot chassis, the ground side of the ac line connects to the internal chassis ground and common-signal grounds of the receiver. As a result, one side of the ac power line carries the 120VAC line voltage while the other side of the ac line connects directly to the chassis and ground. In effect, then, the chassis runs "hot."

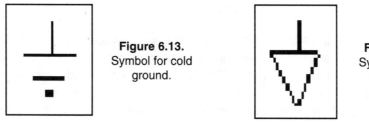

Figure 6.13.
Symbol for cold ground.

Figure 6.14.
Symbol for hot ground.

From a service perspective, the relative lack of protection against plugging the ac line connection incorrectly can result in the connecting of receiver common signal points to the hot side of the ac line. With this type of grounding configuration, any connection between the hot ground and any type of cold ground such as the ground on an antenna system will set up both a shock hazard and the potential for blowing fuses and ruining components in the receiver. Moreover, a technician may risk ruining equipment or receiving an electrical shock when working with a misconnected hot chassis.

To protect against this danger, all electronic devices have a polarized line plug that prevents the incorrect connection to ground. With one blade wider than the other, the polarized plug can be inserted only one way into a standard ac wall outlet. The wider blade connects to the grounded side of the power line that in the wall outlet always corresponds with the white wire. In the receiver, the wider blade always connects to the chassis. Therefore, the use of the polarized plug ensures that the chassis always connects to the grounded side of the ac line.

An *iso-hot chassis* utilizes a combination of the cold and hot grounding systems through the connection of one portion of the receiver to earth ground and the other portion to the ac line ground. Because of the existence of the three different grounding systems, technicians must take special care when using test equipment and when moving from one system to the other. An iso-hot chassis combines the better points of the cold and hot chassis into a different power supply approach and allows the isolated connection of video accessories, such as VCRs, to the receiver. Because of this, the iso-hot design continues to use an isolated transformerless system but also includes a separate cold chassis ground. Any direct connection between the two types of grounding systems will damage the receiver as well as any attached test equipment.

Power Supply Applications for Satellite Receivers

The power supply for a satellite television receiver has several different functions. Regardless of whether the receiver uses an on-screen display or a numeric display, the power supply provides the voltages for the display circuitry. In addition, the power supply provides all the voltages required by the circuitry in the tuner, microprocessor control circuitry, video signal processing circuitry, and audio processing circuitry. As the following sections illustrate, the power supply also provides voltages needed at the low-noise block downconverter, polarizer, and the actuator.

LNB Voltage Injection

A typical LNB requires a dc supply voltage in the 15VDC to 24VDC range. Taken from the power supply of the receiver, the voltage is injected into the coaxial cable connected between the LNB and the receiver. A dc voltage isolation circuit consisting of either capacitors or transformers protects the receiver from the LNB supply voltage.

Polarizer Supply Voltages

Moving to *Figure 6.15*, the polarity control circuit maintains a constant current through the solid-state magnetic polarizer found in the feedhorn. The regulated precise current flowing through the polarizer windings establishes polarization. Because the polarizer is part of an externally mounted feedhorn assembly, the regulation circuits must have the capability to compensate for seasonal temperature changes.

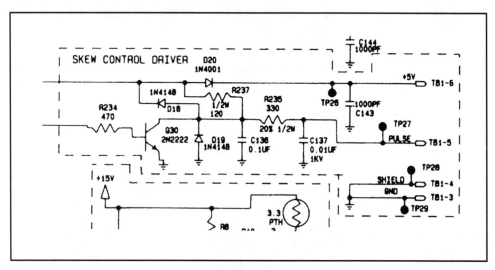

Figure 6.15. Polarizer control circuit.

The direction of the 35 mA of current determines whether vertical or horizontal polarization occurs. In modern receivers, either the tuning microprocessor or a system microprocessor control the current flow. Small memory circuits in the receiver store the current flow direction in relation to the channels and satellites selected by the consumer. In addition to storing the current flow direction, the microprocessor and memory circuits also store the optimum polarization adjustments for each channel.

The fine tuning of the polarizer with these optimum adjustments compensates for subtle differences in the wave format for each channel.

Variations of the design occur through the control of the polarizer through the passing of either zero or 70 mA of current through the polarizer windings. The feeding of current to the servo motor causes a voltage shift to occur and the physical rotation of the polarizer probe through a 90° arc. The duration of the pulses found at the polarizer determine the amount of rotation. The polarity control circuit has three terminals: +5VDC, ground, and pulse.

Receiver Voltage Supplies

Figure 6.16 shows the schematic for the linear power supply found within an integrated receiver. At the left of the diagram, the primary coil of the power transformer connects to the ac voltage input. The secondary coils of the transformer connect to a series of rectifier circuits that exist as part of the +5VDC, +15VDC, +12VDC, +23.3VDC, +22VDC, +6VDC, -12VDC, and -5VDC voltage supplies. Integrated circuits IC1 through IC7 provide regulation for each of the voltage supplies while capacitors C1 through C7 provide filtering.

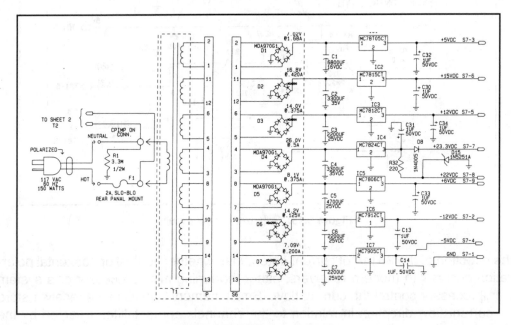

Figure 6.16. IRD power supply.

The positive and negative 5VDC supplies provide voltages for the logic circuits found in the main control board. Logic circuits include the microprocessor and control circuits. In addition, the +5VDC supply works as the voltage supply for the polarotor servo motor. The positive and negative 12VDC supplies provide voltages for the audio and video processing circuits. Audio and video tuning circuits along with the RF modulator utilize the +23.3VDC supply. The +22VDC supply provides the source voltage for the low-noise block downconverter.

Positioner Control and Actuator Voltage Supplies

Figure 6.17 shows a schematic diagram for the actuator control portion of an integrated receiver. Moving from left to right, ICs 8 and 9 translate the position request given by the operator into a command for the actuator. A slow blow fuse and rectifier circuit for the actuator control sits at the top left of the diagram while IC10 provides regulation. To the right of the diagram, ICs 11, 12, 13, and 14 handle data storage for the microprocessor found at IC9. The microprocessor counts pulses and compares the number of pulses with the values found in memory.

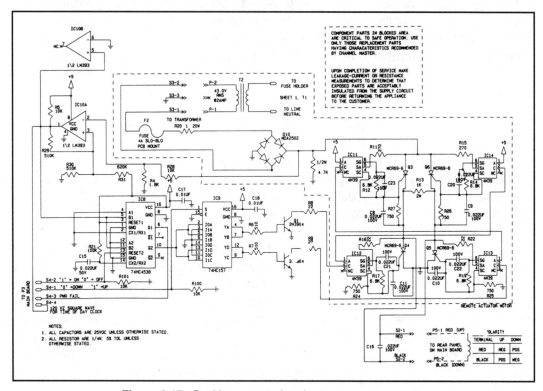

Figure 6.17. Positioner control and actuator power supply.

At the lower right of the diagram, the schematic shows the positioner control circuit connecting to the actuator motor. Referring to the back of the receiver shown in *Figure 6.17*, the red and black connections tie to the actuator motor and, depending on the polarity, cause the motor to turn clockwise or counterclockwise. In turn, the actuator arm retracts and extends.

Common Linear Power Supply Problems

The most frequent type of problem experienced with low-voltage power supplies involves blown fuses or open circuit breakers caused by a shorted rectifier or regulator transistor. Other problems that occur often include leaky or shorted filter capacitors. In many cases, the careful inspection of components in the power supply area can disclose the location of stressed or burnt components. In others, troubleshooting involves carefully checking how the operation of the power supply affects the operation of auxiliary circuits.

When checking a power supply that provides voltages through a voltage divider, disconnect one end of the resistor that supplies power to the load. By disconnecting only one end, you can find if the receiver will remain powered on. While using this type of check, remember that more than one circuit may operate from the same voltage line. As a result, this troubleshooting method may involve disconnecting several circuits in sequence.

Current, voltage, and resistance measurements also provide another method for checking the performance of auxiliary circuits. With the positive end of a diode rectifier disconnected from the circuit, connect one multimeter probe to the disconnected terminal, the other to the circuit connection, and check for the amount current drawn for the given circuit. An unusually high amount of current such as 20 or 30 mA drawn by a circuit indicates that a faulty component is causing an overload.

Lower than normal voltages may show that an overload has occurred in the circuit or that a filter capacitor has opened. In addition, a low voltage may point to a leaky transistor or diode in the regulator circuit. Resistance measurements will verify the open or leaky condition by showing resistance to common ground. Again, disconnect the rectifier diode from the suspected auxiliary circuit and measure the voltage at the source. If the source voltage remains normal even as the receiver shuts down, and the voltage supply is operating normally, then the auxiliary circuit has an open or leaky component. If the source voltage remains at zero volts, a component in the supply circuit has opened or become leaky. As always, you can also use resistance tests to check a suspected diode, transistor, or capacitor.

Filter Capacitor Problems

Many of the problems associated with low voltage power supplies occur because of defective electrolytic filter capacitors. Depending on the symptom, the capacitor may have become shorted, leaky, or open. Defects in filter capacitors occur because of dried electrolytics, broken terminals, and open connections between the capacitor and common ground.

Low-Voltage Regulator Problems

Every type of video receiver has low-voltage regulator circuits that regulate or stabilize dc voltage supplied to stages and circuits throughout the receiver. At face value, each of those symptoms indicates that a defect has occurred within a given section. However, in many cases, the regulator supplying the voltage to the sections has developed a problem.

When you consider regulator circuits, your troubleshooting efforts may involve regulators that use discrete components or regulators using integrated circuits. The first type usually features a large power transistor operating as a variable series-pass resistance between the raw B+ voltage and the regulated B+ voltage. A control circuit varies the bias of the series-pass element to maintain the proper level of regulated B+ voltage at the output. The second type places the regulator circuit within an IC package.

Troubleshooting the discrete regulator circuit involves basic voltage and resistance tests combined with in-circuit and out-of-circuit diode and transistor tests. A very low voltage would indicate that a defect has occurred with the output transistor or the regulator. The same higher-than-normal voltage found at all three terminals of a regulator transistor points to a leaky condition within the transistor. Rotation of the regulated B+ adjustment should cause the output voltage to vary. If the output voltage does not change, the regulator transistor has become defective.

In most cases, you can only use in-circuit tests to find shorted or leaky components within the IC package. As with all IC packages, you can also check the dc voltages and the resistance from pin to ground of each terminal. The voltages found at the IC pins should match voltage readings found on the schematic drawing. Low resistance readings may occur because of leakage within another component connected to the same terminal. To verify leakage within the IC package, disconnect any capacitors, zener diodes, or resistors that parallel the selected IC pin.

Switch Mode Power Supply Problems and Symptoms

Troubleshooting an SMPS problem requires a consistent problem-solving procedure. That procedure should include:

- A preliminary check of the B+ voltage

- A verification of the presence of start-up voltages in a scan-derived power supply

- A verification of the presence of oscillation in a scan-derived power supply

- A check of the SMPS output voltage, and

- A check for regulation

By checking for the presence of B+, we can narrow the search for the problem source from the entire SMPS to the switching device, the bridge rectifier, or the transformer. As shown in the discussion about oscilloscope checks, both the presence and the appearance of the waveform at the switcher are important. After verifying the operation of the SMPS, check the quality of the output voltage. Here, we want to check for proper voltage levels throughout the power supply and for the proper regulation of the output voltage.

Common Problems

Many times, typical problems account for the failure of a switched-mode power supply. Some of those problems involve blown supply fuses, open fusible resistors, high amounts of ripple in one or more output lines, an audible whine with a lower-than-normal voltage at one output, and intermittent power cycling. In many cases, bad solder connections within the SMPS can cause symptoms to appear that mimic component-caused failures.

The blown supply fuse problem may occur because of a shorted switch-mode power transistor or other semiconductors found in the supply. While a fault in the start-up circuit for the supply may cause fusible resistors to open and shut down the supply, the main power supply fuse will not open. When considering ripple in the output lines, check for ripple at the line frequency of 60 Hz or ripple occurring at the switching frequency of 10 kHz or more. A dried filter capacitor connected in the main supply will cause an output line to have a 60 Hz ripple while a dried filter capacitor connected in a specific output line will cause the higher frequency ripple.

The last two symptoms—audible whine with a lower-than-normal voltage and periodic power cycling—involve shorted semiconductors, a fault in the regulator circuitry, a fault in the overvoltage sensing circuitry, or a bad controller. Usually, the failure of a switching transistor is accompanied by the failure of other semiconductors in the circuit. At times, though, a switching transistor will not have the voltage rating needed to withstand the strain caused by the constant on and off switching.

Locating Switching Problems

If the SMPS utilizes a power transistor as a switching device and the power supply fails, always test the transistor for shorted and open junctions. The partial failure of a switching transistor often results from leakage or a change within the operating parameters of the semiconductor. Most new SMPS units rely on either an silicon-controlled rectifier or a metal-oxide silicon field-effect transistor as a switching device. Testing either an SCR or MOSFET requires a multimeter for basic tests such as a shorted condition and additional test equipment for any other tests. When replacing a switching device, always use an exact replacement as recommended by the manufacturer of the SMPS.

Capacitor Problems in Switch-Mode Power Supplies

Any switched-mode power supply design allows a large amount of current to flow through electrolytic capacitors. In some cases, the repeated operation of the SMPS system will cause the capacitors to short internally or develop an intermediate open condition. Under high load conditions, a capacitor may open and then "heal" at line rates. Many times, discoloration or slightly bulged appearance will show that the capacitor has begun to overload.

SMPS Power Cycling Problems

Many SMPS problems involve a dead supply and a sound that either resembles a tweet-tweet-tweet or a flub-flub-flub. In addition, a fault of this nature may cause display LEDs to flash or, with televisions, may allow a partial raster to appear. Most power cycling problems result from a shorted component in the auxiliary power supply. Those components include diodes, capacitors, and SCRs in the overvoltage crowbar circuit.

A failure in the overvoltage sensing circuit will also cause a power cycling problem to occur. If suspecting a failure of this type, check the SCR in the crowbar circuit. Low-power SCRs often operate as control devices in the crowbar circuits used for over-voltage and overcurrent protection. Generally, out-of-circuit tests on the SCRs will disclose any faults. If the power supply fails to energize because of a failure in the protection circuits, the problem may trace back to a shorted SCR.

After checking the component, check for any short circuit conditions on the output lines connected to the SCR. Then, remove the SCR and use a variac to slowly increase the input voltage while monitoring the voltage at the output line. Checking the voltage at this point will show whether the voltage is going past the overvoltage level, if the voltage remains clamped at a low level, or if the voltage stays at the correct level under normal load conditions. A momentary overvoltage spike seen at the turn-on of the receiver will cause the overvoltage circuit to react.

Troubleshooting Auxiliary Power Supply Problems

Often, the shutdown symptom occurs because of a defect in the auxiliary power supply section. With linear power supplies, separate windings of the power trans-former secondary may supply different B+ voltages. Referring back to the previous sections, a switch-mode power supply usually has low-voltage power supply circuits operating from voltages taken from the switch-mode transformer. When we consider either type of auxiliary power supply and the associated circuits, note the number of diodes and electrolytic filter capacitors used in those circuits. A defect in any one of the diodes or capacitors can either cause shutdown or a defect traceable back to several different stages.

Despite the differences seen between linear and switch-mode power supplies, certain patterns remain in place. Each auxiliary power supply extending from a separate winding on a switching transformer will have rectification, filtering, and regulation. In most cases, you will find that a low-voltage power supply consists of a bridge rectifier, a high voltage filter capacitor, and a combination of transistor and zener diode regulation. Knowing that each low voltage circuit probably has these basic parts makes troubleshooting the auxiliary power circuit easier.

Summary

Chapter six introduced you to linear power supplies and the fundamental concepts of rectification, filtering, and regulation. Linear power supplies follow the basic model of input-rectification-filtering-regulation and have the basic purpose of converting an ac line voltage to the required dc voltages.

The chapter provided information about power supplies that use power transformers and transformerless power supplies. As we considered both types of linear power supplies, we found that two different types of grounding systems also exist. You also learned about voltage dividers and bias voltages. The discussion then moved to the difference between cold and hot chassis grounding systems and overvoltage and overcurrent circuits including the popular crowbar circuit.

Chapter six continued the discussion of power supplies with thorough coverage of the switched-mode power supply technology currently used in almost all electronic devices. As the chapter showed, an SMPS uses a rectification-filtering-switching-regulation model and relies on a switching device to pulse voltages obtained from a capacitor. Regulation of the switched-mode voltage may occur through pulse-width modulation, through a combination of pulse-width modulation and flyback regulation, or through the use of a forward switching regulator.

CHAPTER

7

Satellite Receiver Functions and Options

Chapter seven provides an in-depth look at the internal functions and options of a satellite receiver. The chapter begins as the intermediate frequency from the low-noise block downconverter travels into the receiver tuner and concludes with an over-view of various receiver functions. Throughout the chapter, the discussion covers filtering, tuning, frequency conversion, and microprocessor control circuits.

The description of different circuit functions also introduces the reader to a new set of electronic and signal concepts. As the discussion unfolds, the chapter also defines frequency synthesis, phase-locked loops, oscillators, intermediate frequencies, and RF amplifiers. Each point of the discussion links the reader back to materials found in the previous chapters.

Accepting the 70 MHz Signal

Figure 7.1 shows a signal processing block diagram of a typical satellite receiver. The first two blocks of the diagram indicate that the dish antenna gathers the signal while the LNB amplifies and converts the signal into a usable format for the receiver. Moving to the third block of the diagram, the selection of a channel actually selects one transponder from a given satellite. With the conversion of the transponder frequency to an RF carrier, this stage provides a further downconversion of the signal into a new intermediate frequency or IF signal.

Filtering and Amplifying the IF Signal

In *Figure 7.1*, a surface acoustic wave, or SAW, filter acts as a tuned circuit, establishes the intermediate frequency bandwidth, and provides waveshaping. The SAW filter is constructed from a piezoelectric substrate that allows the device to function like a crystal in that it resonates at specific frequencies. However, because the device also has two sets of different length metal electrodes extending across the substrate that allow the two transfers of energy, the SAW filter works as a double transducer.

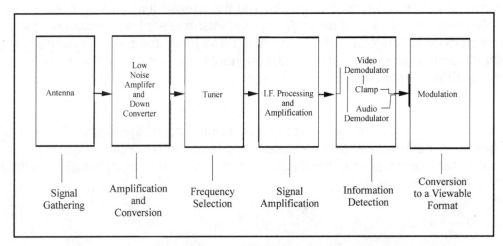

Figure 7.1. Signal processing block diagram of a typical satellite receiver.

Figure 7.2 shows a basic diagram of a SAW filter. In the figure, the IF signal appears at the input array of the SAW filter as an ac signal. The array is called an amplitude weighted array because the varied length fingers provide a sharp cutoff for a particular response. Because of the piezoelectric effect, the twisting or bending of the crystal substrate converts the electrical signal into a mechanical vibration called an acoustic wave. The vibration occurs at the signal frequency and creates waves that travel across the surface of the SAW filter to its output array.

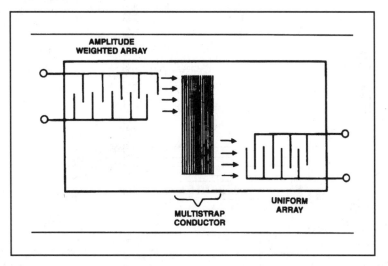

Figure 7.2. Diagram of a SAW filter.

Given specific crystal and array parameters, the filter uses the vibrations to establish the desired bandwidth and frequency response characteristics. Various types of crystal materials provide a better response and allow less signal loss at desired IF signals. The weighting of the input and output arrays can effect selectivity and bandwidth. Increasing the number of array fingers limits bandwidth while decreasing the number of fingers broadens the bandwidth. At the output of the SAW filter, a uniform array—consisting of a specified number of uniform fingers—maintains the bandwidth within the desired 6 MHz range.

Tuning Systems

Figure 7.3 provides an interconnect diagram for an integrated receiver that includes a decoder module. Matching the diagram to the block diagram shown in *Figure 7.1*, the intermediate frequency from the LNB travels to the tuner. Processing of the signals that output from the tuner occurs within the main board of the receiver while a keyboard allows customer input into the receiver. At the bottom right of the dia-

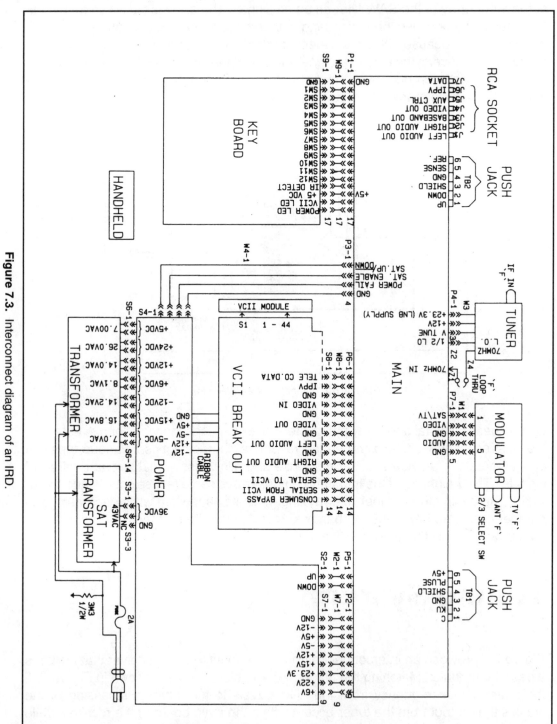

Figure 7.3. Interconnect diagram of an IRD.

gram, the power board provides regulated dc voltages for the receiver circuits, the positioner control, and the actuator.

Using Digital Circuits to Manipulate Frequencies

Almost all modern communications devices rely on some type of frequency division, multiplication, or synthesis. In satellite receivers, frequency synthesis appears in tuner circuits as a method for selecting channels; and in control circuits for setting the proper operating frequencies. Frequency synthesis involves the use of digital devices such as counters, dividers, prescalers, encoders, decoders, and comparators. The type of frequency synthesis used in consumer electronics also introduces voltage-controlled oscillators and phase-locked loops.

Frequency Division and Multiplication

Frequency division involves producing an output signal that has a fractional relationship—such as 1/2, 1/3, or 1/10—to the input signal. A *frequency multiplier* consists of 1) a time-varying circuit that introduces harmonics at the output along with the fundamental frequency; and 2) a resonant circuit tuned to the desired output frequency. The resonant circuit passes only the desired output frequency to the load while rejecting other frequencies including the fundamental frequency.

As an example, a frequency tripler uses a 1 MHz oscillator to convert dc power from a power supply line into a 1 MHz sine wave. When the sine wave feeds a nonlinear amplifier, the amplifier symmetrically distorts the signal so that the positive and negative peaks flatten. With this flattening, the amplifier introduces odd harmonics into the signal.

As a result, the output signal consists of the 1 MHz fundamental frequency along with a sequence of odd harmonic frequencies. Starting with the third harmonic, each harmonic is progressively less than the fundamental. At the output, an LC tank circuit tuned to 3 MHz passes the 3 MHz signal while rejecting the fundamental and all other harmonic frequencies.

Frequency Synthesis

Frequency synthesis is a method of digitally generating a single desired, highly accurate, sinusoidal frequency from the range of a highly stable master reference oscillator. The desired frequency corresponds with a precise function of subharmonic and/ or harmonic relationships found in the reference oscillator frequency. With this, a frequency synthesizer translates the performance of the reference oscillator into useful frequencies. When designs employ several reference oscillators, the possible number of output frequencies exceed the number of oscillators.

Most low-cost frequency synthesis circuits use frequency division to produce the desired frequency. The frequency division occurs through the use of a counter because of the ability of the device to function over a wide bandwidth. Frequency synthesis accomplished through this method can only produce an output frequency that has a lower value than the input frequency. Because crystal oscillators have a top frequency of 200 MHz, frequency synthesizers using frequency division are capable of working at frequencies less than 100 MHz.

Other frequency synthesizer circuits rely on frequency multiplication. With this technique, an oscillating signal passes through a diode, transistor, or varactor diode. Then, the nonlinear semiconductor elements produce harmonics of the original signal. After that, the harmonically-rich signal passes through a sharp, narrowband filter that attenuates any undesired harmonics.

Anotherfrequency synthesis method involves the mixing of frequencies. A mixer multiplies frequencies and generates a signal that contains both the sum and the difference of the two input frequencies. The circuit uses one frequency as the desired frequency and discards the other through filtering.

Frequency Synthesizer Operation

Figure 7.4 illustrates the operation of a basic frequency synthesizer. In the figure, the master reference oscillator has two outputs. Two divide-by-ten circuits divide the oscillator frequency by ten and then subdivide those dividends by ten. As a result, two accurate subharmonics—a tenth and a hundredth—of the master oscillator frequency exist.

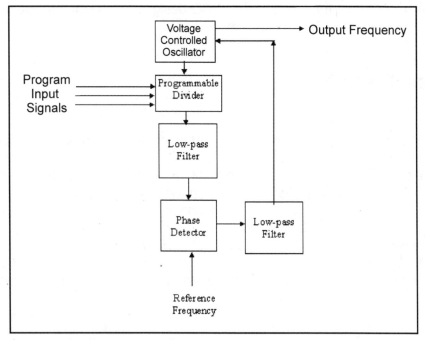

Figure 7.4. Block diagram of a basic frequency synthesizer.

Within this frequency synthesis system, the output of a stable reference oscillator divides into precise subharmonics. Then, some type of discrete or integrated switching arrangement would select the proper harmonics, add the frequencies, and output the signal through a bandpass filter. The filter would allow only the desired frequency to pass and has an extremely precise frequency at the output.

The locking portion of a frequency synthesis circuit depends on the precise measurement, division, and comparison of two input frequencies. While the tuner oscillator supplies an input signal, a reference oscillator provides a known and stable reference frequency. When we begin to look at the frequency synthesis circuit as a whole, dividers and a prescaler measure the input pulses as a series of binary numbers and divide the numbers so that the output is a usable low frequency. Each off and on of the pulse represents a binary number.

Voltage-Controlled Oscillators

Indirect frequency synthesis circuits usually rely on a *voltage-controlled oscillator*, or *VCO* to generate a frequency. With the VCO, a square-wave output frequency has an inversely proportional relationship with the input voltage. Therefore, a low input control voltage produces a high output frequency and a high input control voltage produces a low output frequency. Most VCO designs rely on the precise frequency relationships given by a 555 astable timer.

Phase-Locked Loops

A *phase-locked loop*, or PLL, contains a voltage-controlled oscillator, prescaler and divider circuits, a comparator and a quartz crystal, and provides a low-cost alternative to frequency synthesis. The design of the phase-locked loop provides the most stable operating conditions. As *Figure 7.5* shows, the PLL receives an input signal and then compares that signal with the feedback of an internal clock signal generated by the VCO. Given the tuning provided by the PLL, the VCO oscillates at a frequency where the two divided signals are equal and adjusts the feedback signal so that it matches the reference signal applied to the phase detector in both frequency and phase. As a result, the internal and external clock signals synchronize.

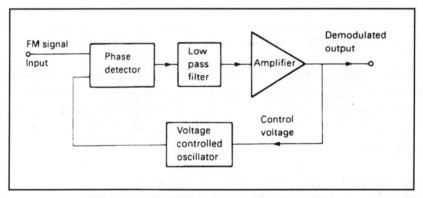

Figure 7.5. Block diagram of a phase-lock loop.

Many PLL designs rely on a "charge pump" consisting of inverters, switches, and a passive RC low-pass filter. Looking again at the figure, the input signal from the first frequency divider clock enters a phase detector. In this example, the phase detector consists of a set of buffers and D-type flip-flops. The phase detector compares the clock input with a feedback signal from the VCO. The frequency divider portion of the PLL may utilize a basic digital counter or something as complex as a single-sideband modulator and translates the input and output frequencies to usable levels.

PLL Advantages and Disadvantages

The phase detectors used in PLLs offer the advantage of easy implementation because of the use of digital techniques. In addition, phase-locked loops offer good noise characteristics because of the conversion of noise to phase noise. Moreover, phase-locked loops offer extremely good frequency locking. The key disadvantage of PLLs is the length of time needed to switch from one output frequency to another output frequency.

Oscillator and Mixer Stages

Tuner oscillators produce a sine-wave frequency that has a much higher level than the incoming frequency synthesized signal. With the output of the oscillator stage connected to the input of the mixer stage, the high frequency, high amplitude oscillator output is injected into the mixer. This change occurs through the heterodyning, or mixing, of the high frequency oscillator output signal and the radio frequency signal associated with the selected channel. With this operation, the tuner works with a signal that changes with the selection of a channel and produces an intermediate frequency signal that never changes.

As a result, the stages following the tuner do not need to retune every time the customer selects a new channel. The mixer output signal contains the sum and the difference of the IF output signal from the LNB and oscillator output signal. The difference frequency becomes the second IF output signal and contains baseband sound and picture carriers.

Selecting a Channel

As with all receivers, a satellite television receiver allows the consumer to select channels. With 24 transponders assigned to a C-band satellite, up to 32 transponders assigned to a Ku-band satellite, and 50 transponders assigned to a DBS satellite, the consumer has a broad selection of programming channel choices. Therefore, the receiver must have a control system for channel selection. When the consumer selects a channel, the receiver applies a dc voltage to the block downconverter.

Satellite signals rely on the block downconversion of the entire 500 MHz block of satellite television frequencies to a lower block of frequencies. With the satellite television frequencies occupying the 3700 MHz to 4200 MHz range, different manufacturers have selected different frequency blocks for downconversion. In the past, those frequency blocks have included the 270 to 770 MHz range, the 940 to 1440 MHz range, and the 1000 to 1500 MHz range. Modern satellite receivers rely on the 940 to 1440 MHz range. *Figure 7.6* shows a basic diagram of the block downconversion system.

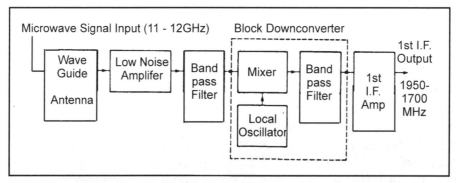

Figure 7.6. Basic diagram of a block downconversion.

The Federal Communications Commission has assigned precise frequency channels for the transmission of satellite television signals. Depending on the age, manufacturer, and design of the tuner, the receiver can cover the entire 500 MHz band from 3700 to 4200 MHz. In contrast, other tuner designs allow the selection of only the 24 available channels, spaced every 20 MHz from 3720 MHz to 4180 MHz.

Figure 7.7 provides a schematic diagram of an electronic controlled tuner control system used within a satellite receiver. At first glance, the schematic seems extremely complex because of the need for the exact tuning of channels, automatic fine tuning, automatic channel search capabilities, and easy pushbutton control. By dividing the complete design of the electronic tuning control system into sections, we can gain a better understanding about how the control system functions and how to apply standard troubleshooting methods. The four sections are:

- A dedicated microprocessor and system memory section to issue and store operating commands

- Customer Preference Controls and Displays

- A frequency synthesis section designed for exact channel selection and continuous frequency control, and

- An electronic bandswitching system

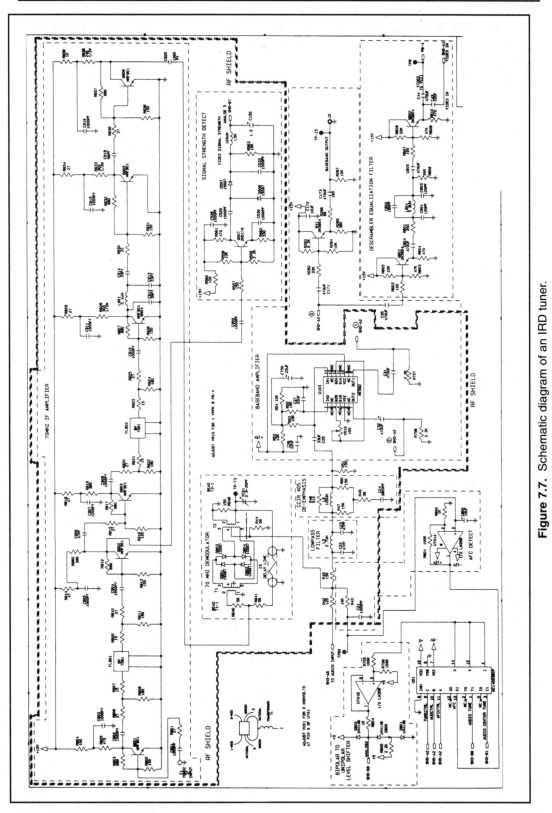

Figure 7.7. Schematic diagram of an IRD tuner.

To accomplish these tasks, the microprocessor interfaces with from either a keypad or remote control receiver; the system memory, and logic circuitry and follows a simple input/process/output routine. As *Figure 7.7* shows, data travels from either the keyboard—which contains customer and channel selection controls, the remote control interface, or a memory section into the microprocessor, and into an input section.

Referring to *Figure 7.8*, U1, the microprocessor responds by storing data in the tuner memory and the sending of the channel selection information to the frequency synthesis circuitry. After leaving the CPU, the information travels through U4, a latch, and into the memory, U2, or along an output path to either the channel display, the control circuit at U8 and U10, or the frequency synthesis circuit. The control circuit handles signal voltages for the AFC, tuner, audio, positioner, and video processor circuits.

Data travels to the central processing unit or CPU in the form of binary numbers. The binary number system provides a convenient method for performing low-level instructions and for both temporarily placing information into and retrieving information from the random-access memory or RAM of the tuner. Located at U2 and U3, the read-only memory or ROM—holds preset, permanent instructions for the microprocessor. Whether configured as *read-only* or *random-access memory*, the memory devices store both channel selection information and most-often needed instruction sets. The instructions take the form of a pre-programmed set that the microprocessor follows when prompted by some type of signal.

Along with using memory for storing data, the microprocessor also temporarily stores data in a series of registers of latches. When the microprocessor issues an instruction routine, a sequence in that routine may use the data group during a routine operation or may use the group to latch desired conditions. Because the register offers only temporary storage, the system discards the stored data after completing the operation.

Every tuner system requires some method for allowing the customer to select a channel, control the tuning, and see the channel display. Because of the evolution of video products, tuner control systems have undergone many changes. In the recent past, tuner control systems have evolved from the mechanical tuners to a hybrid approach that used potentiometers and varactor diodes and finally to designs based on frequency synthesis. With frequency synthesis, electronically controlled tuners combine a control assembly with a tuning system, control processor, memory, frequency synthesizer control units, and different types of keyboard input.

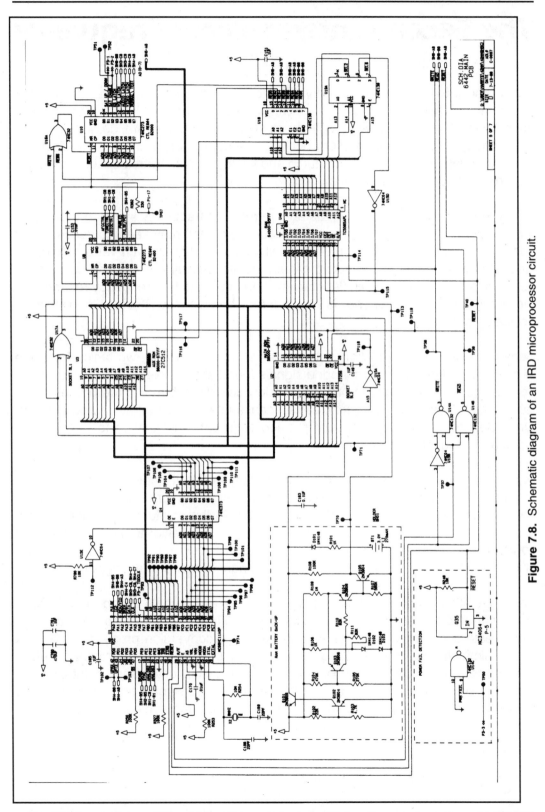

Figure 7.8. Schematic diagram of an IRD microprocessor circuit.

The Second Intermediate Frequency

Block four of the diagram featured in *Figure 7.1* shows that amplification of the IF signal must occur before the receiver can translate the signal into a format recognized by the television receiver. With the next stage in the diagram, demodulation, or the removal of usable information, occurs. Demodulation of the intermediate frequency signal provides baseband audio and video signals. With a baseband signal, no further frequency conversions occur.

The mixer stage of the tuner outputs an intermediate frequency signal to the IF section. Although the terms "intermediate frequency signal" may seem to describe one specific frequency, the mixer output actually contains signals with various frequencies. Decreasing the frequency to the intermediate level allows an increase in amplifier gain, provides better amplifier stability, and improves the frequency response of the circuit. The lower frequency is the difference between the RF signal and the local oscillator frequency.

Because of this, the IF section of the receiver must complete two tasks. First, the section must eliminate the unneeded signals found at the mixer output. Also, the IF portion uses tuned circuits to provide not only selectivity but also the proper signal ratios. Any unwanted IF frequencies can cause interference in the reproduced picture. Second, the IF section must provide sufficient amplification for the desired video and sound IF signals.

Figure 7.9 shows the schematic diagram for the second IF tuner control as well as the receiver control circuitry for the polarotor. Within the tuner control circuit, data travels along three lines into the controller while 3.2 MHz crystal establishes the local oscillator frequency. Pin 4 connects the circuit to the receiver tuner. The polarotor control circuit separates into the pulse and sense circuit at the top right of the diagram and the skew control driver circuit at the right center of the diagram.

Bandpass filters within the final IF stage eliminate any signals outside the 28 MHz bandwidth and set the channel bandwidth. After filtering occurs, amplifiers in the IF stage restore signal strength lost during the down conversion process. A limiting circuit eliminates sparklies in the video signal by producing a square wave output from a sine wave input signal and increasing the efficiency of the demodulator circuit. After the signals are shaped to eliminate the undesired signals and then amplified, demodulation of the signals begins the development of the audio and video signals.

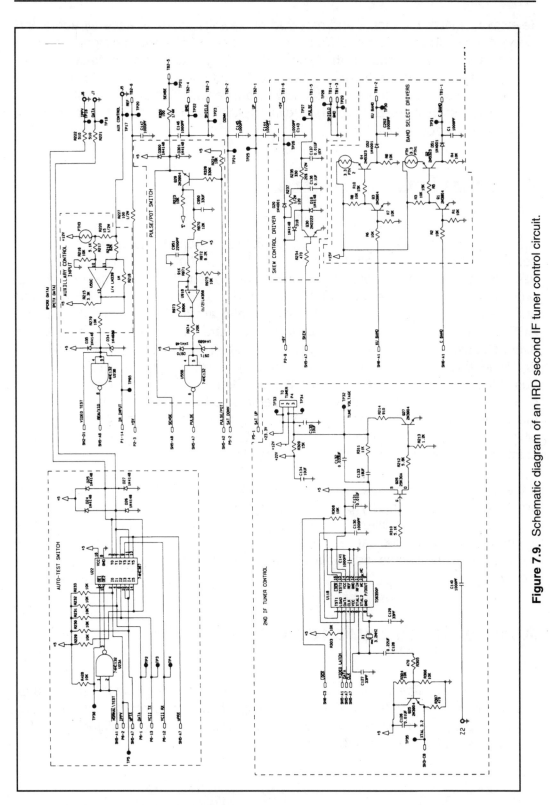

Figure 7.9. Schematic diagram of an IRD second IF tuner control circuit.

Television signals broadcast and received by satellite systems have all audio and video information added through frequency modulation. The satellite signal carries a greater amount of information than that seen with typical broadcast television signals. This information includes a number of audio subcarriers used for multi-language broadcasts, stereo signals, and news broadcasts. Because the satellite broadcast signal also includes baseband video information, we can refer to it as a composite baseband video signal.

Automatic Gain Control

Any type of communications equipment includes an automatic gain control circuit for controlling the gain of a signal amplifier. *AGC*, or *automatic gain control*, circuitry uses signals from the IF section to monitor and correct the overall gain of signals processed within the tuner and the IF section. In video products, the RF amplifier and the IF amplifiers connect to an AGC line and are forward biased at all times. AGC circuits take advantage of the forward bias characteristics of the amplifiers and either change or maintain amplifier gain by adjusting the operating conditions of the amplifier. Because of this, the dc control voltage produced by the AGC circuit may push the amplifier towards either saturation or cutoff. The AGC circuit works as part of a closed feedback loop that includes a detector, a filter circuit, a feedback path to the amplifier, and the amplifier stage. When the input signal varies in amplitude, a dc correction voltage feeds from the AGC circuit to the amplifier and maintains a constant amplifier gain by controlling the amplifier forward bias.

If the input signal amplitude increases, the AGC circuit prevents the output of the amplifier from increasing by producing a higher dc bias control voltage. The increased bias at the amplifier reduces the gain of the amplifier stage. A decrease in the input signal amplitude causes the AGC circuit to produce a lower dc bias control voltage. Here, the decreased bias at the amplifier causes the amplifier stage gain to increase.

Automatic Frequency Control

AFT, or *automatic fine tuning*, circuitry uses signals from one of the IF amplifiers to monitor and correct any IF signal changes by applying a dc correction voltage to the tuner. Properly operating tuner and IF sections require the precise alignment of the signals processed throughout the circuits.

Detecting or Demodulating the Signal

The intermediate frequency output from the block down converter feeds through the IF section of the receiver and directly into the demodulator. *Figure 7.10* shows the frequency spectrum of the baseband signal. Video information fits within the 30 Hz to 4.2 MHz range of the frequencies. Because of the characteristics of frequency modulation, two different types of circuits may provide demodulation.

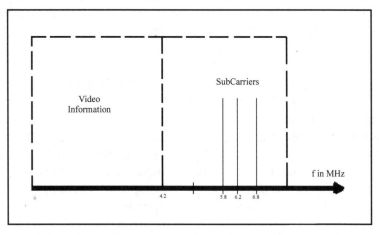

Figure 7.10. Frequency spectrum of the baseband signal.

With the first, a circuit called a frequency discriminator combines with a limiting circuit to ensure that the amplitude of the signals remains constant. In those circuits, changes in frequency cause a change in the output voltage from the sound carrier. The second and most popular type of demodulation circuit involves the use of a phase-locked loop.

Supplying a Usable Video and Audio Signal

The video bandwidth for analog satellite television broadcasts can range from 30 MHz to just below 18 MHz. Narrowing the bandwidth below the 18 MHz range degrades the picture quality while allowing the transmission of more channels per transponder. With the narrower bandwidth, the signal becomes more susceptible to noise interference. The same problem with narrow bandwidth also occurs at the receiver. At the receiver, the video bandwidth ranges from 28 MHz to 36 MHz. A bandwidth below 28 MHz often results in the loss of picture detail, variance in color shading, and the presence of sparklies in saturated red and green colors.

Video signals transmitted to satellites have higher frequency components that are emphasized. With this, a simple electronic circuit called a pre-emphasis circuit amplifies the high frequency components of the video signal at the point of transmission. Pre-emphasis becomes necessary because the noise output at the FM demodulator has greater amplitude at higher frequencies. At the receiver, a de-emphasis circuit lowers the amplitude of the higher video frequencies and creates a flat noise response at the output of the FM demodulator and removes most of the audio subcarrier signal.

Given this de-emphasis, the video signal has an appearance similar to that shown in *Figure 7.11*. In the figure, the video signal rides on a 30 Hz triangular waveform called the energy dispersal waveform. The shape of the waveform prevents terrestrial interference from interrupting the satellite communications. If the transmission of the RF carrier lost the video information, the energy dispersal waveform prevents the transponder carrier frequency from spreading across the bandwidth allocated to the transponder. Instead, the use of the energy dispersal waveform limits the spread to approximately 2 MHz of the transponder bandwidth.

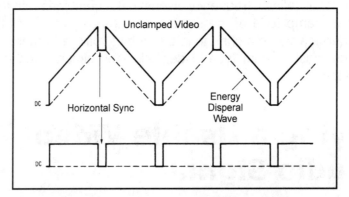

Figure 7.11. Video signal after de-emphasis.

Without the energy dispersal waveform, the unmodulated transponder carrier frequency would cause large-scale interference. In addition, the video signal would consist of only the horizontal synchronization pulses and the 3.58 MHz color subcarrier. Concentrating the signal energy in this form would cause interference with terrestrial microwave link transmissions.

At the receiver, a clamping circuit removes the energy dispersal waveform so that the receiver can produce a suitable video signal. The *clamping circuit* illustrated in *Figure 7.11* holds the sync tips of the composite video signal to a fixed voltage level. Clamping the video signal to the fixed level restores the signal to the form originally seen at the video detector output. The lack of clamping at this point would cause the picture to flicker at a 30 Hz rate.

Baseband Processing

The baseband audio and video signals consist of 15 Hz to 15 kHz signals for audio and 30 Hz to 4.2 MHz signals for video. While the baseband format signals may work well for video monitors, the use of the signals for a television requires an additional stage. The modulator seen in the schematic diagram shown in *Figure 7.12* adds an RF carrier to the baseband audio and video signals. Usually, the frequency of the RF carrier coincides with the frequency needed for either channel three or four within the television receiver.

Moving from the tuner/demodulator module, the baseband video signal has an output of 250 mV peak to peak and includes the video information, the energy dispersal waveform, and the audio subcarriers. At the output of the baseband processing circuitry, the signal splits into the video processing stage, the sound demodulator stage, and the external baseband stage. Some decoders, such as the MAC and D-MAC decoders, utilize the 1 VDC peak-to-peak external baseband signal before de-emphasis and clamping occur.

At the video processing stage, the de-emphasis circuitry cleans the raw baseband signal and feeds the signal into a video amplifier stage. As a result, the signal increases to the 1VDC p-p level found with the standard composite video waveform. Moving to the output of the video processing stage, we find the baseband composite video signal seen in *Figure 7.13*.

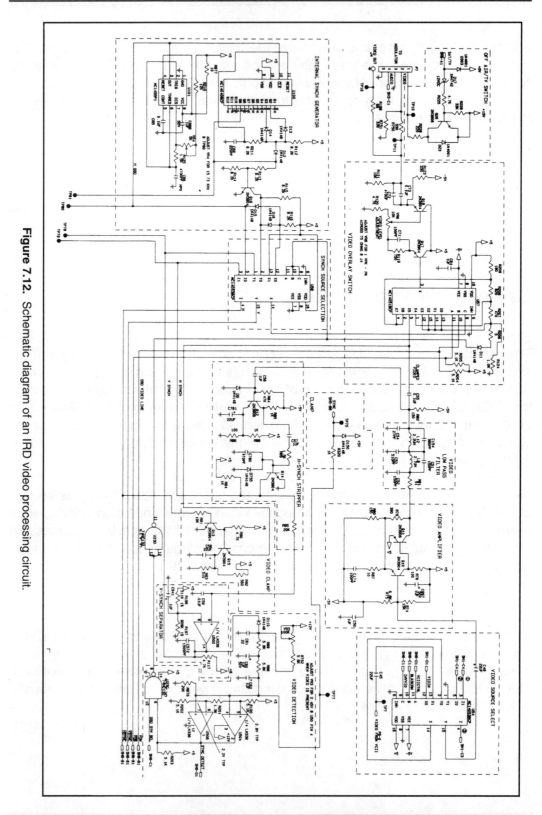

Figure 7.12. Schematic diagram of an IRD video processing circuit.

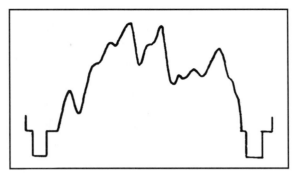

Figure 7.13. Baseband composite video signal.

Remodulation

While it may seem more efficient to simply input the 70 MHz IF signal to the television cable connector, the use of frequency modulation for the audio and video signals rules this option out. Television receivers use frequency modulation for audio signals and amplitude modulation for video signals. Because of this, the satellite receiver requires an additional modulation circuit that remodulates the baseband signals after the initial demodulation. Re-modulating the signals formats the composite video signal so that it matches the requirements of the television receivers.

Satellite television receivers use the carriers for either VHF channel 3 or VHF channel 4 for the remodulation process. Channel 3 has a frequency band of 60 to 66 MHz; a video carrier of 61.25 MHz; and an audio carrier of 65.75 MHz. Channel 4 has a frequency band of 66 to 72 MHz, a video carrier of 67.25 MHz, and an audio carrier of 71.75 MHz. The consumer has the option of selecting either channel 3 or channel 4 through a switch located on the back of the receiver.

RF Amplification

Electronic systems take advantage of information carried within RF signals to produce sound and pictures. The *RF amplifier* provides both selectivity and amplification while increasing the voltage level of an IF signal applied to its input. Given its flat bandpass, the stage should equally amplify all passing frequencies and rejects any signals that lie outside the bandpass. Also, RF amplifiers should have a good signal-to-noise ratio. With these two characteristics in mind, the RF amplifier stage design strengthens desired signals while canceling internally- or externally-generated noise. Internal noise may arise from the mixer or from the conduction of some semiconductor components. External noise is a product of electromagnetic field generated by appliances, power lines, and unfiltered automobile ignition systems.

Receiver Signal-to-Noise Ratio

One of the key measures for receiver quality is the signal-to-noise ratio found at the output of the receiver. With this, the signal level in decibels must have a higher amplitude than the noise level in decibels. An acceptable signal-to-noise level always allows the reception of a clear picture. *Chart 7.1* compares signal-to-noise ratios and picture quality standards.

Signal-to-Noise Ratio	Picture Quality
46.6 and above	Excellent
42.3 to 46.6	Good
38.0 to 42.3	Fair
33.2 to 38.0	Marginal
29.3 to 33.2	Poor

Chart 7.1. Comparison of Signal-to-Noise Ratios and picture quality standards.

Receiver Threshold

By definition, the threshold of a receiver shows that the device operates at standard input and output levels. However, two different methods for measuring threshold exist. Dynamic threshold measurements rely on the Carrier-to-Noise-Ratio, or CN/R, for the receiver and measure performance under operating conditions. Measured in decibels, the Carrier-to-Noise Ratio of a receiver illustrates the relationship between maximum signal strength traveling into the receiver and noise from internal and external sources. Manufacturers measure the threshold point in decibels. When the Carrier-to-Noise Ratio of the receiver yields a level below the threshold point, performance deteriorates in the form of noisy video.

The measurement of the static threshold point occurs when no useful information is received at the device and involves the video Signal-to-Noise Ratio. At the receiver, nothing but the unmodulated carrier appears at the input. With this, static threshold measures the amount of continuous quieting versus the Carrier to Noise Ratio. Techniques involving the use of phase-locked loops and filters can extend the threshold level of a receiver to lower limits.

Receiver Options

Most—if not all—satellite receivers contain advanced microprocessors and memory storage circuits that provide advanced functions for viewers. Many receivers arrive with factory installed software programs that automatically perform the tasks required to receive the available satellites and satellite TV services. As an example, many receivers store pre-programmed locations and tuning parameters for all of the available satellite TV services in a memory circuit.

Along with pre-programmed instruction sets, receivers also arrive with remote controls that provide access to large number of functions. Combinations of key functions work much like a computer keyboard while the television doubles as a computer display for installation and programming menus. With the use of those menus, a viewer can customize an integrated receiver to meet specific viewing needs.

As opposed to beginning of satellite television when a customer might have a separate receiver, antenna positioner, and descrambler, modern receivers integrate all functions into one device. An integrated receiver/descrambler, or IRD, contains the descrambler module that can decode encrypted subscription-only programming. When selecting an IRD, always ensure that the decoder module is compatible with the desired programming service.

Providing Monophonic and Stereo Sound

When transmitting monophonic sound, a satellite system uses only a single subcarrier. As an example, a satellite equipped with 24 transponders could use the same subcarrier frequency for the audio signal that accompanies a televised image. We can refer to the single subcarrier as subcarrier A. The transmission of stereophonic requires the use of two subcarriers—one for each sound channel. However, the subcarriers do not remain the same for each program. We can refer to the subcarrier pair as A + B.

While a satellite may use one or a pair of audio subcarriers, it may also utilize more than one pair. As an example, a transponder on Satcom F3 may use the 5.40 MHz and 5.44 MHz subcarriers for the audio of one program and another subcarrier pair at 5.50 MHz and 5.76 MHz for another audio program. The use of different stereo modes also affects the subcarriers. Discrete stereo requires two active subcarriers that have identical signal levels. One subcarrier carriers the left audio channel signals while the other carries the right channel audio signals. Matrix stereo also utilizes two active subcarriers. However, one subcarrier has a higher level than the other. The lower level subcarrier has the Left minus Right subcarrier difference frequency while the higher level subcarrier is the Left plus Right subcarrier sum frequency.

Figure 7.14 shows the schematic diagram for a stereo processor circuit found in a satellite receiver. At the bottom left of the diagram, the left and right audio signals travel from the Videocipher module into pins 4 and 8 of IC8, a preamplifier/stereo processor. IC9 provides additional amplification and processing while IC7 functions as an MPEG audio decoder. IC10 provides additional output amplification for the monophonic and stereo signals.

Inverting the Video Signal

Many satellite receivers will have a video inversion feature. The modulation of the video signal occurs through a process called negative modulation. With this, the most negative portions of the composite video signal shown in *Figure 7.15* produce the blacker parts of the signal while the synchronizing and blanking pulses have the highest modulating signal voltages. Video inversion establishes a condition where the blanking and synchronizing signals produce the brightest light on the television screen. The change in polarity for those signals also means that the reproduced picture will not have synchronization. As a result, the picture will tear vertically and horizontally.

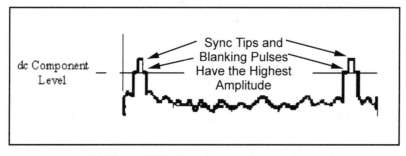

Figure 7.15. Composite video signal.

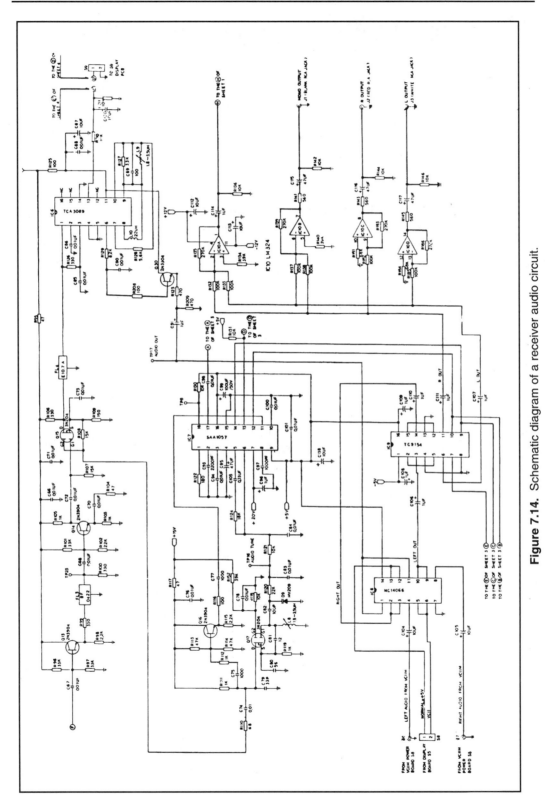

Figure 7.14. Schematic diagram of a receiver audio circuit.

Providing C- and KU-Band Signal Reception

Video inversion becomes a greater factor when we consider receivers that accept and process both C- and Ku-band signals. Any receiver manufactured for dual-band reception utilizes video inversion. This occurs because C-band downconverters utilize high-side injection while Ku-band downconverters utilize low-side injection. As a result, the Ku-band signals appear inverted with respect to C-band signals and vice versa.

Both C-band and Ku-band include frequencies that fit within the 950 MHz to 1450 MHz bandwidth. However, C-band frequencies range downward from 1450 MHz to 950 MHz while Ku-band frequencies range upward from 950 MHz to 1450 MHz. When comparing C-band and Ku-band signals, we find that the channel order and the video signal are reversed.

Customer Preference Displays and Controls

Control of any tuner begins with the customer interface. In the past, a consumer working with a mechanical tuner used a knob attached to a shaft to select and fine tune satellite channels. With electronic tuners, the customer interface and the control system begins with some type of keypad or remote control device. In addition, the customer interface includes either a channel display based on light-emitting diodes or LEDs or an onscreen display.

Obviously, channel selection and tuning begin when the customer uses either a pushbutton, a keypad, or a remote control device to start instruction sets into motion. In addition to working with receiver controls, the customer also needs some method for finding which channel is selected. While the older mechanical tuners relied on a light bulb that illuminated the numbers on a rotating wheel, modern tuners use either displays based on light-emitting diodes or onscreen displays.

LED Channel Displays

Figure 7.16 shows a block diagram for a channel display using light-emitting diodes and data from a microprocessor. When the customer chooses a channel by either touching the front-panel keyboard or using the remote control, the microprocessor senses the key closure. Then, the processor sends data, clock, and enable signals to the channel display circuit through three output lines. The display circuitry—which includes a combination of latches, buffers, and a shift register—decodes and stores the serial data. The proper segments of the channel display are driven to light by the data. In addition, three voltage sources—+5.4VDC for the display circuitry, +3.2VDC for the LED, and +4.3VDC for the LED segment brightness—power the display assembly.

Onscreen Channel and Menu Displays

Many satellite television receivers use onscreen displays to show everything from the time, volume settings, channels, and current channel to customer preferences and menu selections. In all cases, a character generator enclosed in an integrated circuit produces alphanumeric characters which can be positioned in different areas of the screen. Generally though, the current channel number will display in the upper right screen and the time will display in the lower left corner.

The display circuit takes advantage of the red, blue, and green output signals found at the chrominance/luminance control integrated circuit. Each selection of a new channel causes the onscreen display integrated circuit to send a blanking signal to the character blanking transistor Q1. The transistor shorts the output signals from the chroma/luminance IC to ground while the data from the onscreen display IC output goes to the RGB buffer transistors.

Remote Control Functions

Figure 7.17A shows the schematic for an infrared remote transmitter. The transmitter sends a 14 bit signal containing channel and volume information to the remote preamplifier circuit shown in *Figure 7.17B*. During operation, the remote preamplifier applies the 14 bit signal to pin 9 of the microprocessor. When the first bit of the data string reaches pin 9, the digital level at the pin goes high and the microprocessor releases digital low enable pulses at its pin 8. The presence of the enable pulses allows the remote preamplifier to send the remainder of the data to the microprocessor.

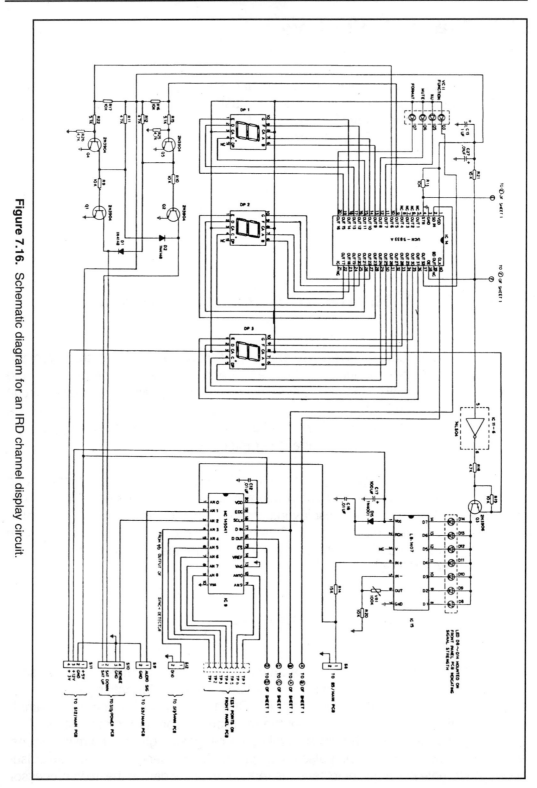

Figure 7.16. Schematic diagram for an IRD channel display circuit.

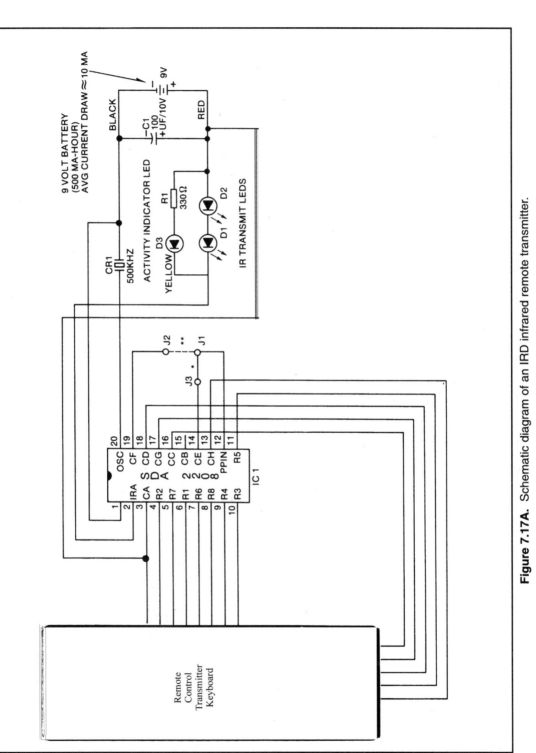

Figure 7.17A. Schematic diagram of an IRD infrared remote transmitter.

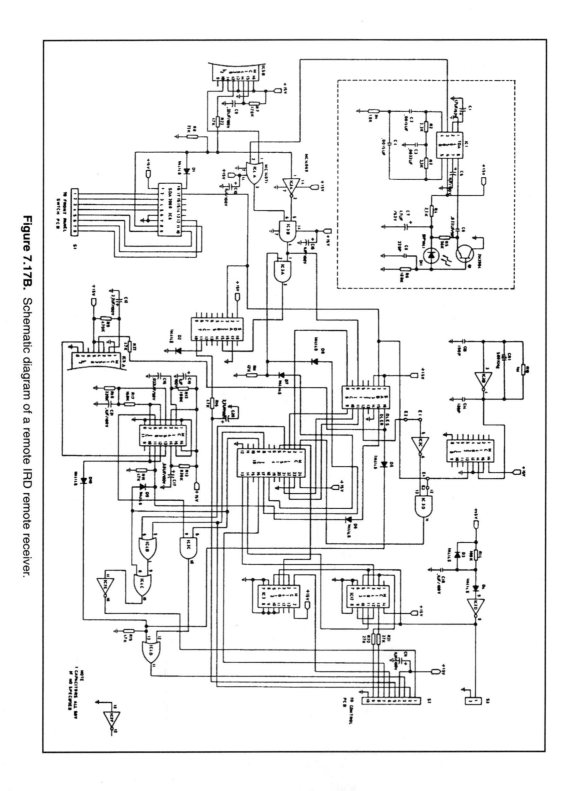

Figure 7.17B. Schematic diagram of a remote IRD remote receiver.

Parental Lockout

Parental lockout allows a viewer to designate certain satellite channels as off-limits to children. Shown in *Figure 7.18*, an onscreen menu allows the viewer to program the receiver to accept the channels only after the submission of a password. With some integrated receivers, parental lockout may also prevent children from changing the tuning parameters of the system.

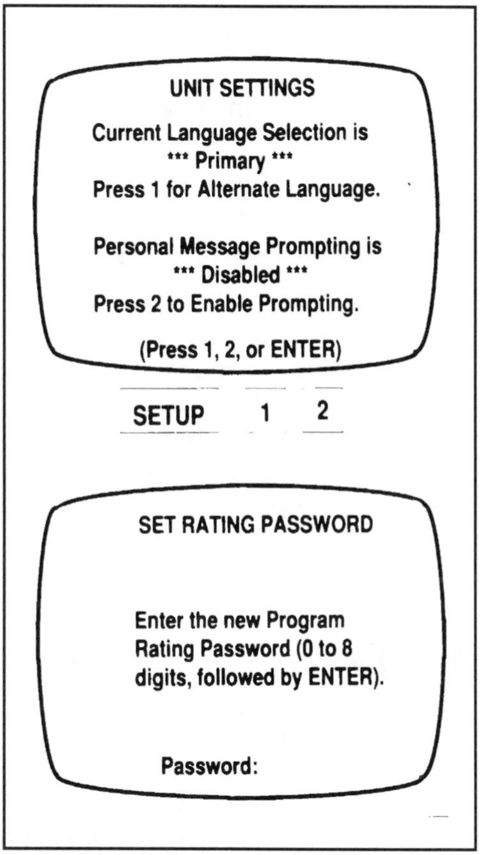

UNIT SETTINGS

Current Language Selection is
*** Primary ***
Press 1 for Alternate Language.

Personal Message Prompting is
*** Disabled ***
Press 2 to Enable Prompting.

(Press 1, 2, or ENTER)

SETUP   1   2

SET RATING PASSWORD

Enter the new Program
Rating Password (0 to 8
digits, followed by ENTER).

Password:

Figure 7.18. Onscreen menu for parental lockout.

Automatic Peaking

Most satellite television receivers include an automatic peaking function that allows the viewer to fine tune the system performance if the quality of received signals has deteriorated. As an example, degradation may occur if high winds have pushed the dish a fraction out of alignment. Automatic peaking measures the strength of the incoming signal while moving the dish back and forth in small increments. In addition, automatic peaking tunes the channel frequency and polarization settings.

With automatic peaking engaged, the receiver automatically selects settings for peak signal reception. The viewer can accept those settings and store the new settings into the receiver memory or return to settings previously stored in the memory for the desired satellite. Automatic peaking usually will not work with receivers that operate with smaller dishes because the receiver may easily lock onto a signal from an adjacent satellite.

Automatic VCR Time Setting

Many receivers include a function where the viewer can synchronize the receiver internal clock with the internal clock of a VCR and set the receiver so that it tunes to a specific satellite and channel at a selected time. Most receivers allow the programming of multiple events over a period of time ranging from two to four weeks. As a result, the viewer can easily tape programs broadcast through satellite television providers for later viewing.

Favorite Channels

Most mid- to high-price range receivers allow viewers to create a customized list of favorite audio and video services. With the favorite channel option, a view does not need to remember the name of any satellite or the channel number for a given satellite service. Instead, the viewer assigns the desired program to a favorite channel list and selects the service with one touch of a remote control button.

Summary

Chapters six and seven combine to provide an "up-close" look at integrated satellite receivers. While chapter six emphasized the power supply circuitry found within the receiver, chapter seven considered signal processing circuitry. As with chapter six, chapter seven used the description of the circuitry functions to define a series of electronic concepts and terms. In addition, the chapter relied on a series of actual schematic diagrams from different satellite receivers while describing the circuits and signals.

The information found in chapters six and seven also applies to the discussions given in chapters eight, nine, and ten. While chapters eight and nine cover direct broadcast satellite systems, digital signals, and digital television, chapter ten provides valuable information about descramblers. As a whole, the information guides you through satellite television technologies.

CHAPTER

8

Digital Satellite Broadcasts and Digital Television

As with most electronic systems, satellite television reception systems have evolved from relying on analog signals to the utilization of digital signals. With digital satellite signals, computer technology converts the analog video and frequency modulated audio signals to a series of ones and zeroes before transmission. The transmission of the converted signals then involves the compression of data. Because of conversion and compression, direct broadcast systems can place more programming channels on individual transponders and provide clearer pictures at the home.

Chapter eight introduces DBS systems by first explaining the transmission of Ku-band signals. As the discussion progresses, the chapter defines how direct broadcast satellites function and describes how digitized satellite signals differ from the standard analog signals in terms of bandwidth and signal strength. As with other chapters in this text, the chapter makes a transition from theory to practice by walking readers through a typical installation of a DBS system.

The chapter concludes with a discussion about digital television signals. As the chapter continues with this discussion, it defines different digital signal formats. With this, the chapter also describes the impact of high-definition television and other digital formats on consumers and the DBS industry.

Ku-Band Signals

Until now, the chapters in this book have concentrated on C-band satellite television reception systems. However, the transmission and broadcast of television signals at Ku-band frequencies rather than C-band offers several benefits. At the higher Ku-band frequencies, the bandwidth of the signal allows the use of smaller, easier-to-install receiving dishes. Moreover, signals transmitted at Ku-band frequencies have better signal quality.

In North America, consumers use C-band systems, C/Ku-band systems, and direct broadcast satellite systems. In Europe, consumers have made Ku-band systems that rely on an offset-style dish that includes a low-noise block feed, or LNBF. Varying the supply voltage for the LNBF from 14VDC to 18VDC allows the device to switch between the vertical and horizontal polarities. In the United States, several digital satellite services providers use the same method to switch between left-hand and right-hand circular polarization.

Ku-Band Frequencies

In North America, the Ku-Band frequencies separate into the 11.7 to 12.2 GHz Fixed Satellite Service band, the 12.2 to 12.7 GHz direct broadcast satellite band, and the Broadcast Satellite Services band. While some DBS services utilize the FSS band, several television stations also transmit video feeds and data on the 11.7 to 12.2 GHz frequencies. The DBS band remains designated only for direct to home satellite signal applications. LNBs used to amplify direct broadcast satellite signals have a local oscillator frequency of 10.75 GHz and an intermediate frequency ranging between 950 to 1450 MHz.

Digital Satellite Signals

The growing popularity of DBS systems extends from the capability to directly broadcast many channels of cable-quality programming from satellites to dishes that measure from 18 to 26 inches in diameter. At the writing of this book, more than one in ten households in the United States have a DBS dish and subscribe to digital programming. The systems work through the digital compression of standard analog television signals and the subsequent delivery of those 200 or more channels from a single transponder located on a geosynchronous satellite.

Direct Broadcast Satellites

With launch dates beginning December 17, 1993, the generation of satellites operating in the 12.2 to 12.7 GHz portion of the Ku band has become the support backbone for digital communications services. Each of the three satellites—DBS-1, DBS-2, and DBS-3—has 16 transponders that provide the high output of 120 watts needed send usable signals to eighteen inch dish antennas. The transponders have a radiated broadcast power that ranges between 48 and 53 decibel-watts; the satellite as a whole generating 4300 decibel-watts.

The satellites orbit at 101 degrees West Longitude and have a 0.5-degree separation. From this location, the satellites can cover the entire United States. This location corresponds to a North/South line passing through western Nebraska. Using that line as a reference point, the satellites appear due South when viewed from the central United States. From the Eastern United States, the satellites appear West of South while citizens living in the Western United States would see the satellites as placed East of South.

With traditional C-band and Ku-band satellite signal transmissions, the overlapping of frequencies caused interference between dedicated microwave links and the Fixed Satellite Service operators. Because DBS satellites utilize the high power portion of the Ku-band for downlink frequencies and 17 GHz for uplink frequencies, the transmission of DBS signals avoids the frequency overlap problem.

Direct broadcast satellites use a combination of two receiving and two transmitting antennas that allow the reception of the maximum number of stations without tracking the satellite. Although the DBS satellites set in geosynchronous orbits, the consistent reception of high quality signals occurs through the use of shaped receive and transmit reflectors based on a uniform antenna structure. As a result of this structure and the referencing of the antennas to the satellite beacon, the satellite has highly accurate pointing, decreased weight, and an increase in both output power and the effective isotropic radiated power.

Antennas on the satellites utilize both left-hand and right-hand circular polarization to transmit the signal energy to receiving stations. Odd numbered channels use the left-hand format while even numbered channels utilize the right-hand format. While the simultaneous use of left-hand and right-hand polarization increases the complexity of the home receiving systems, the transmission of signals in both formats allows the broadcast of more 24 MHz-wide channels in the same frequency band. In addition, the use of circular polarization and a 5.16 MHz guard band minimizes any cross-channel interference.

Traveling Wave Tube Amplifiers, or TWTAs, provide the operating power for the sixteen transponders found on direct broadcast satellites. In addition to operating with analog and digital modulation formats, the 120-watt amplifiers can work in low- and high-power modes. When operating in the high mode, the power from two amplifiers combines for the operation of one transponder. With this, a satellite could offer a full complement of 120-watt transponders or half the number of 240-watt transponders. Currently, the DBS-1 satellite transmits in the low mode while the DBS-2 and DBS-3 satellites operate in the high mode. The three satellites deliver 53 to 58 decibel-watts of power to the respective coverage areas.

In the United States, the Federal Communications Commission has allocated eight orbital positions at the equator for U.S.-owned high power direct broadcast services. Because the orbital positions serve the continental United States, the satellites are often referred to as full-conus slots. Each one of the positions will have a maximum of 32 broadcast transponders.

Because of the coverage given by the full-conus slots at 101 degrees, 110 degrees, and 119 degrees, providers attempt to have as much programming as possible at those positions. DIRECTV has a license for 32 transponders at 101 degrees, three at 110, and eleven at 119, while EchoStar has 27 transponders at 110 degrees, 21 at 119, and several others that serve only the east coast or west coast. With this, the providers have potential access to anywhere from 300-400 total channels.

Any limit to the number of channels compressed onto each transponder depends on desired image quality, the degree of allowable visible artifacts, the frame rate of the source material, the amount of movement in the source material, and the amount of error correction overhead. In addition, the number of channels also depends on the amount of additional information included within the compression such as program guide data, conditional access data, and error correction data used to precisely reconstruct the original signals.

Analog-to-Digital Signal Conversion

Current over-the-air broadcasts and most cable systems rely on the NTSC, or National Television Systems Committee, standard for the transmission and reception of television signals. The analog standard utilizes 525 horizontal scanning lines, interlaced scanning, the transmission of separate luminance and chrominance signals, and relies on a 60 Hz frame rate. Of the 525 scanning lines, only 483 are visible while the remaining lines are used for interval timing or other functions. The bandwidth for those signals covers 4.2 MHz.

Although the transmission of signals occurs in a digital format, direct broadcast system providers continue to use NTSC programming and display the results on conventional television sets. To accomplish this, the providers digitize the analog signal before transmission. Decoders at the viewer's homes convert the digital signal back into a standard NTSC signal for display. The preciseness of the digital transmission and the compression scheme used for the signals allows the received signal to exactly match the source signal. Interruptions in the reception of the high frequency digital signals may occur, however, because of atmospheric conditions such as heavy rain. In addition, some signals may remain susceptible to digital artifacts that seem to freeze a portion of a picture or allow small colored boxes to appear.

The number of channels compressed to a transponder depends on the content of the program. Four to five channels that contain programming with many fast-moving small objects such as a basketball game can compress to a transponder before significant digital artifacts begin to appear. In comparison, six to seven channels that contain programming with mostly large slow-moving images can compress to a single transponder. As an example, movies film at a very constant, non-interlaced 24 frames per second and contain less source material. Movies are filmed at 24 frames per second rather than 30 for video so they contain less source material. In some instances, eight to nine channels carrying movie programming can compress a single transponder and maintain acceptable quality.

In addition to programming content, the amount of actual picture and sound data sent across a transponder also depends on the amount of error correction needed for the signal. Forward error correction restores data that may have been lost during transmission. The use of more powerful satellites or multiple satellite transponders configured to broadcast in the same frequency can reduce the error correction overhead and allow the carrying of additional program data. With multiple satellites in orbital locations, the power on each transponder frequency increases and reduces the need for error correction.

Digital Satellite Services

Not surprisingly, the broadcast market for digital satellite signals is extremely competitive. At this time, DIRECTV and EchoStar dominate the direct broadcast market with their respective DIRECTV® and DISH® services that offer comparable content and price. Co-developed by Hughes Electronics DIRECTTV and Thomson Consumer Electronics, the DSS, or Digital Satellite System offers both the satellite reception system and programming services.

Digital Satellite Equipment

A direct broadcast satellite system includes the reflector, mount, LNBF, receiver, decoder, and remote control. Given the size of the receiver, many customers purchase the systems as kits and follow self-installation guides. DBS systems either integrate the decoder as part of the receiver or utilize a separate set-top box decoder. The remote control provides access to installation and setup routines as well as a variety of customer controls.

DBS Reflector Size

In a previous section of this chapter, we found that DBS satellites have increasing amounts of radiated power. However, the true measure of effectiveness remains with the amount of power reaching the dish. The amount of available broadcast power reaching the dish affects the required size of the reflector. This occurs because of the antenna gain and beamwidth factors described in chapter three. As an example, a Ku-band satellite transponder transmitting 60 watts of power will require a 24-inch reflector for the optimum reception of the signals.

Because DBS satellites transmit at higher power levels, reflector sizes have dropped to 18 inches. An 18" x 20" reflector offers an approximate gain of 34 dB at 12.5 GHz and a half-power beamwidth of 3.5 degrees. The other factor affecting reflector size and the quality of the received signals is the transmission data rate. DBS transponders operate at a data rate of 40 megabits per second. While the transmission of programming that has relatively little action requires only 2 megabits per second of video compression, action scenes require higher bit rates that may produce coding errors during compression and decompression.

Because Ku-band signals occur at high frequencies and have smaller wavelengths, the signals will not penetrate objects. The short wavelength causes the signal to reflect from almost any solid material and refract easily. As shown in *Table 8.1*, atmospheric conditions such as rainfall can severely attenuate Ku-band signals. Every three decibels of loss cuts the received power in half. If your DBS system sets at a low altitude, the system will experience more rainfade because the distance increases the amount of rain in the path between the reflector and the satellite. Larger DBS reflectors that measure from 24 inches to 36 inches in diameter collect more signal and reduce the loss due to rainfade.

Rate of Rainfall	Signal Loss
up to 1/8 inch per hour	~ 0.01 decibel
1/8 to 1/2 inch per hour	~ 1 decibel
½ to 1 inch per hour	~ 2.5 decibels
1 to 4 inches per hour	~ 12 decibels
4 to 10 inches per hour	~ 40 decibels

Table 8.1. Signal loss due to rainfade.

DBS Decoders

The decoder found within a DBS receiver or as part of a set-top box first decodes the digitized signal and then converts the data into usable video and audio signals. Because a decoder decodes only one channel, viewers wishing to watch different programming at the same time will need individual recorders. DIRECTV and the DISH network do not utilize compatible decoding systems. When shopping for a DBS system, the customer will need to ensure that the decoder functions with the desired programming.

Installing a Direct Broadcast System

Before beginning the installation of a DBS system, always complete a site survey similar to that shown in *Figure 8.1*. Generally, the electronics store or retail store selling the system will provide the azimuth and elevation settings for DBS systems in their area. Moreover, many DBS manufacturer web sites provide calculators that accept geographic locations as input and provide the azimuth and elevation angles for the area.

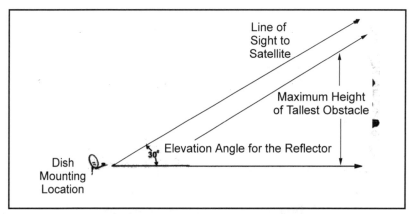

Figure 8.1. Line of sight survey example.

Using a compass and standing at least one foot away from any metal buildings, hold the compass so that the needle can align with magnetic North. Take a compass and the azimuth and elevation angles to the location for the satellite dish. All compasses divide into 360 degrees with zero degrees representing North, 90 degrees representing East, 180 degrees representing South, and 270 degrees representing West. With this in mind, find the azimuth angle for your area on the compass and mark that angle at the reflector location.

Once installed, the reflector must have an unobstructed view of the satellite. The elevation angle measures from the ground to the sky from any location. When considering the elevation angle, 90 degrees represents vertical, zero degrees represent horizontal, and 45 degrees represents the halfway point. As you plan for the location of the reflector and mast, use your arm to estimate the elevation angle for your location and point to the satellite. The elevation angle will vary from satellite to satellite. When checking for obstructions, consider the growth patterns of any trees located near the installation site. The clear view found during one season may give way to a tree full of leaves or a neighbor's tree that has grown taller. In addition, always consider the necessity for cleaning snow or other debris from the reflector. With this, mount the reflector in an easily accessible location.

As opposed to C-band systems, the small size of DBS system hardware also allows the mounting of the reflector and mast to the side of a building or to a ground pole. Before drilling any mounting holes for the reflector and mast, check for the proximity of water pipes, security system cabling, or electrical cabling to the mounting site. If setting a ground pole for the system, check with the utility companies to ensure that no water or electrical lines intersect with the proposed ground pole location. Although most cities take a relaxed stance towards zoning and DBS systems, neighborhood covenants may restrict the location of the reflector. While checking zoning laws, also check for any neighborhood or building restrictions regarding the placement or use of DBS systems.

After verifying that the location will have a clear line of sight from the reflector to the satellite, check the routing of the RG-6 coaxial cable that connects from the LNBF to the wiring hub or the receiver. The reflector should mount as close to the receiver as possible to ensure that no line loss occurs. RG-6 coaxial cable will carry high frequency signals up to 100 feet with no line loss while any distances that exceed 100 feet require the use of a line amplifier. When installing the cable, take care to ensure that the cable does not have any sharp bends or kinks and that the mounting clips do not pierce any portion of the cable.

A wiring hub is a central location for all cable connections and becomes especially important for multi-receiver installations. Even though the receiver may arrive with the single LNBF shown in *Figure 8.2*, always install two RG-6 cables. With this, the installation accommodates a system upgrade to a multi-receiver setup and the use of the dual LNBF shown in *Figure 8.3*. When installing a wiring hub, place the hub in an easily accessible location that allows expansion or repairs.

Figure 8.2. Simple LNBF.

Figure 8.3.
Dual LNBF.

Installing the Reflector and Mast

Always assemble the reflector, mast, and LNBF before climbing a ladder to the mount location. Referring to *Figure 8.4*, fasten the reflector to the support using four flat head mounting bolts and lock nuts. Although the bolts and lock nuts should fasten securely, avoid overtightening the fasteners and harming the reflector.

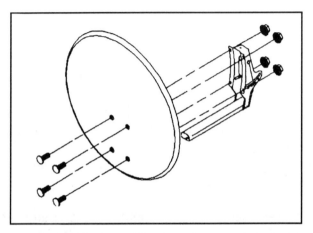

Figure 8.4. Assembling the reflector and mast.
Courtesy of the Dish Network.

Moving to *Figure 8.5*, always use a level to ensure that the mast remains at true vertical. After assembling the reflector and support arm and verifying that the mast is vertical, tilt the support bracket to the approximate elevation angle for the desired satellite. The support arm for the reflector will either align elevation marks with points as shown in *Figure 8.6* or align a colored edge of the mast with a slot on the bracket as shown in *Figure 8.7*.

Installing the LNBF

Whether working with a C-band, Ku-band, or DBS system, always treat the low-noise amplifier with care. Referring to *Figure 8.8*, place the LNBF onto the support arm and gently slide the device into place until it seats firmly. The LNBF should attach to the support arm with one mounting screw.

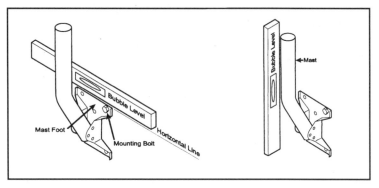

Figure 8.5. Using a level to verify the placement of the mast.

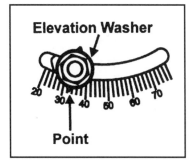

Figure 8.6. Aligning the support arm with pre-cut elevation marks.

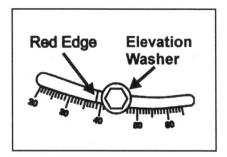

Figure 8.7. Alternate alignment method for the support arm.

Figure 8.8. Installing the LNBF onto the support arm.

After mounting the LNBF, thread the RG-6 coaxial cable into the bottom and out the top portion of the mast. As *Figure 8.9* shows, thread the cable into the lower portion of the support arm and out the upper end of the arm while extending approximately three inches of cable from the end of the support. In addition, allow the cable to have some slack for the purposes of easily removing the amplifier and for a drip line. Use a cable clip to fasten the cable to the mast.

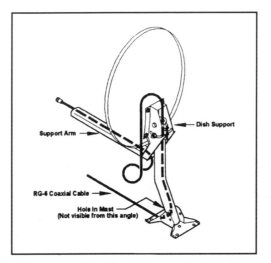

Figure 8.9. Installing the transmission table.

Connecting the Receiver

Figure 8.10 shows the back of a typical DBS receiver. With the reflector, support arm, mast, and LNBF assembled and mounted at the permanent location, connect the RG-6 cable to the Satellite IN connector of the receiver. Then connect another coaxial cable between the TV SIGNAL OUT connector on the receiver and the signal input connector on the television.

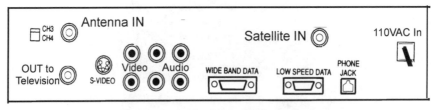

Figure 8.10. Back of a DBS receiver.

Connecting a Phone Line To The Receiver

All DBS receivers require an active connection to a telephone line. The connection allows pay-per-view and other service providers to communicate directly with the receiver and establish the proper billing. Referring to *Figure 8.10*, a standard telephone line connects to the back panel connection labeled Telephone Jack. With the line connected, use the remote control keys to program the receiver with the proper phone number.

To do this, select the menu, main menu, system setup, installation, and installation and setup functions and menus. From there, select the telephone system option for the receiver and the telephone system setup option. The menu system will ask for a selection between a touch tone and rotary/pulse phone at the phone type list. With all options, pressing the select key on the remote control will select the desired option while the save key saves the selection. The final request from the telephone system setup menu involves the selection of a one- or two-digit code at the prefix option.

Aligning the System

All DBS receivers use an internal installation program for the final setup and alignment of the system. After powering the system, use the remote control to set the receiver to the SAT mode so that the commands control the receiver. Using the control to access the Main Menu provides access to the System Setup and Installation menus. At the Installation menu, select the Installation and Setup menu and the Point/Dish Signal option. At this point, most DBS receivers provide an option for typing in the zip code for the specific location.

From there, the receiver displays the azimuth and elevation angles for the desired satellite. The up/down/left/right arrow buttons on the remote control allow the selection of different satellite name and location options. At the reflector, turn the reflector and mast so that the LNBF support arm aligns with the satellite.

Electronic Alignment of the Receiver and Reflector

Along with the physical alignment of the reflector, mast, and support arm, DBS systems also require the electronic alignment of the receiver and reflector. As with the telephone system setup, the alignment of receiver and reflector takes advantage of the remote control and internal menu functions. The alignment begins with the selection of the main menu and then the system setup and installation options. With the installation option displayed, the next step in the process involves the selection of the installation and setup menu and the point dish/signal option.

Most DBS receivers provide a satellite selection option at this point. With this, the customer has the option of selecting a satellite and specific azimuth and elevation angles. In addition, customers may input the zip code for their location and allow the receiver to select the best satellite for viewing. With the reflector aimed at the satellite, information downloads from the satellite to the receiver.

Part of the installation involves observing the signal strength bar found on the front panel of the receiver. A red signal strength bar displaying the word "unlocked" and a beeping signal tone signify that the receiver has not received the strong signal needed for good reception. Aside from the "unlocked" error message, the signal strength bar can also display a "wrong satellite found" message. With this message, the receiver indicates the presence of a strong signal but a signal from a different satellite than selected at the opening menu. A green signal bar showing the word "locked" indicates that the reflector is aimed at the desired satellite and that a strong signal exists. As opposed to the beeping signal tone, the locking onto the correct satellite results in a steady tone that rises with increasing signal strength.

The next part of the installation involves slightly moving the reflector while observing the signal strength meter and the picture. Even though the picture may seem clear, continue to adjust the reflector for the strongest possible signal. With the strongest signal, picture quality will not suffer as much during bad atmospheric conditions. When adjusting the reflector, move the assembly from side to side first and then up and down. Always tighten the support arm, mast, and elevation angle bolts after every adjustment. Once you have adjusted the reflector for the strongest possible signal, firmly tighten the mast and support arms bolts and mark the locations on the mast and support arm with permanent ink.

The last step in the alignment and installation process involves downloading information from the satellite to the receiver and decoder. For this part of the process, exit from the point dish and signal strength menu and switch the remote control from SAT to View. Before downloading the information, the receiver will respond by displaying a command such as attention and then asking if the installation is complete. Once the viewer selects the "yes" option with the up/down/left/right controls of the remote control, the receiver will show a warning message indicating that the download will begin. Using the select button will start the process of downloading any update or configuration data to the receiver from the satellite. At the end of the download process, most receivers will display an "acquiring satellite signal" message.

Digital Television

Until now, we have discussed the conversion of traditional analog signals to a digital format. By 2006, the traditional method for transmitting and receiving analog television signals will give way to digital television. The move to digital television signals will affect the direct broadcast satellite industry in terms of bandwidth and equipment.

Pixels and Refresh Rates

All this leads us to some basic discussions. When we talk about video displays, the ability of the display to show a clear image is defined through a constant called dot pitch. In any display, a *pixel* consists of three individual red, blue and green dots. *Dot pitch* is the distance between the center points of adjacent horizontal pixels on the CRT screen. Most advertisements for video display monitors will list the dot pitch measurement in millimeters. Any video display that has a smaller distance between pixels will have a higher possible resolution. A lower dot pitch number—such as .28—shows that the dots are closer together than those seen with a .31 dot pitch.

Each line that results from the vertical and horizontal scanning of the CRT electron beam yields a set number of pixels. The longer horizontal lines will have more pixels than the shorter vertical lines. If the specifications of a monitor list a resolution of 640 x 480 pixels, the horizontal scan lines show 640 pixels while the vertical line shows 480 pixels. Multiplying the two figures gives us the total amount of pixels that the raster will display. In this case, the total number of pixels is 307200. Since the number of pixels depends on the deflection signals, varying the horizontal scan rate also varies the number of displayable pixels.

Televisions and older computer monitor computer designs have a horizontal frequency of 15,734 kHz. Newer computer monitor standards and the new high definition television standard use horizontal sync signals of 21.80 kHz, 31.50 kHz and 35 kHz. By retaining the 60 Hz vertical scan rate and increasing the horizontal scan rate, more horizontal lines become squeezed into the vertical cycle.

An increased number of horizontal lines further improves the clarity produced by the video monitor. Information display monitors also use higher picture bandwidths than television receivers. In other words, the monitor turns its display pixels off and on quicker than a television receiver does. We know that television receivers have a bandwidth of 4.5 MHz. Information display monitors have a bandwidth of 35 MHz or higher. The higher bandwidth allows the monitor to display more pixels during one horizontal scan. Without the needed bandwidth, a monitor is limited in the resolution that it can provide.

Refresh rate defines the rate at which a screen image is redrawn, shows how many frames are scanned per second, and is the vertical scanning rate. Because CRTs form images in frames, the amount of refresh rate coincides with the amount of flickering seen on the screen. A refresh rate between 60 and 75 Hz with a refresh rate of 75 Hz becoming commonplace. *Table 8.1* lists bandwidth measurements in combination with resolution, the number of pixels, and horizontal sync rates:

60 Hz Refresh Rate			
Resolution	**# of Pixels per Screen**	**Bandwidth**	**Horizontal Sync Rate**
800 x 600	480000	28.8MHz	36kHz
1024 x 768	786432	47.2MHz	46.1kHz
1152 x 900	1036800	62.2MHz	54kHz
1280 x 1024	1310720	78.6MHz	61.4kHz
66 Hz Refresh Rate			
800 x 600	480000	31.7MHz	39kHz
1024 x 768	786432	51.9MHz	50.7kHz
1152 x 900	1036800	68.4MHz	59.4kHz
1280 x 1024	1310720	86.5MHz	67.6kHz
72 Hz Refresh Rate			
800 x 600	480000	34.6MHz	43.2kHz
1024 x 768	786432	56.6MHz	55.3kHz
1152 x 900	1036800	74.7MHz	64.8kHz
1280 x 1024	1310720	94.4Mhz	73.7kHz

Table 8.1. Pixels, bandwidth, and horizontal sync rates.

Interlaced Scanning

When television was first introduced, the *NTSC*, or *National Television Systems Committee*, selected interlaced scanning as a standard for broadcast signals because of the limited bandwidth available for delivering picture information. *Interlaced scanning* is a process where electron guns draw only half the horizontal lines with each pass. With one pass, the guns draw all odd lines while the next pass draws all even lines. As a result, one complete frame of information is created for every two fields scanned. With field generated every 1/60th of a second, the human eye cannot discern the scanning motion. To compensate for any possible flicker, manufacturers of interlaced scanning displays choose phosphors that have a higher decay time.

Because interlaced scanning refreshes only half the lines at one time, it can display twice as many lines per cycle. Thus, the display technique provides an inexpensive method for yielding more resolution. Interlaced scanning has a relatively slow trace and retrace time that affects the ability of a display to show animations and video graphics.

Non-Interlaced Scanning

Interlaced scanning has two problems. Because of the higher resolution, any amount of flicker caused by screen phosphor decay would be noticeable and distracting. With all the individual dots displayed, some will dim as others become illuminated. In addition, the scanning lines in an interlaced scanning display are visible. If a person stands too close to a display device, each line of information can be seen as it displays on the screen. For that reason, the optimal viewing distance for an interlaced display is always listed as 4.5 or 6 times the height of the display screen. At this distance, scanning lines seem to merge together and create the illusion that one complete image is displayed. However, with larger display devices—such as projection televisions—the scanning lines are more noticeable.

To counter the flicker and scanning line problems, computer displays and the new HDTV standard use *non-interlaced refresh* or *progressive scanning*. With progressive scanning, every line of information on the display is scanned by the electron gun at each pass across the panel. The technique enhances the vertical resolution of the display while allowing the viewer to sit closer to the display. Viewing distances with progressive scanning shorten to 2.5 times the height of the display.

Changing Television Signal Transmission Standards

The current television broadcasting and reception standard was established by the NTSC in 1940. That standard utilizes 525 horizontal scanning lines, interlaced scanning, the transmission of separate luminance and chrominance signals, and relies on a 60 Hz frame rate. Of the 525 scanning lines, only 483 are visible while the remaining lines are used for interval timing or other functions. The bandwidth for those signals covers 4.2 MHz.

Although many nations use the NTSC standard, most European nations rely on another standard called Phase Alternate Lines, or PAL signal that relies on a 50 Hz frame rate, uses a color subcarrier frequency of 4.43 MHz, and has 626 scanning lines. The alternate standard surfaced because of detectable shifts in the color subcarrier phase of the NTSC. Still another standard—developed by the French and known as SECAM, or SEquential Coleur Avec Memoire, is used in the former East Block European countries. The introduction of each broadcast standard also introduced incompatibilities between each system. For example, a SECAM system cannot display a PAL broadcast image because of differences between broadcast equipment. NTSC systems cannot display PAL broadcasts because of the difference in the frame frequencies.

In the early 1980's, the Japan Broadcasting Corporation, or NHK, proposed the MUSE HDTV interlaced system that would use 1,125 scan lines and introduced it as a possible world standard. With this proposal, NHK established a goal of high-definition television playing on a wide screen format. At the request of broadcasters concerned about America's role in establishing the new technology, the FCC established a rule making committee called the Advisory Committee on Advanced Television Service, or ATSC. In addition, the FCC decided that new HDTV signals would be broadcast on currently unusable channels and that broadcasters would be temporarily assigned a second channel for the transition to HDTV.

During the early 1990's, three competing high-definition television design teams agreed to combine their efforts and produce a standard, high-quality product. The three design teams—working under the direction of AT&T and Zenith Electronics, the General Instrument Corporation and the Massachusetts Institute of Technology, and Philips

Consumer Electronics, Thomson Consumer Electronics and the David Sarnoff Research Center—have formed the *Digital HDTV Grand Alliance.*

The high definition television standard produced by the Grand Alliance establishes a technological framework for the merging of broadcast, cable, telecommunications, and computer technologies. Not surprisingly, the introduction of a high definition television standard affects both the transmission system and the receiver design for modern video receivers. With *HDTV* the amount of luminance definition doubles both horizontally and vertically. An HDTV system provides four times as many pixels as the older NTSC system. In addition, the wider aspect ratio given by HDTV systems establishes more visual information for the viewer. Specifically, the high definition system yields increased vertical definition through the use of 1125 lines in the scanning pattern. The system provides additional video detail through the application of video bandwidth five times that seen with the conventional NTSC system.

While the NTSC system offers an *aspect ratio*—or the ratio of picture width to picture height—of 4:3, the increased picture width given by the HDTV system establishes an aspect ratio of 16:9. Because of this, the viewer gains the capability to receive almost six times more information. Therefore, high-definition televisions have a place in industrial, information capture, storage and retrieval, educational, medical, and cultural applications. With each of those applications, HDTV provides the picture quality needed for teleconferencing, training, and product promotion. *Figure 8.11* compares the NTSC standard picture size with the HDTV picture size.

The HDTV 1125/60 Standard

During the planning and development of the HDTV system, the design team chose to use 1125 scanning lines with a picture refresh rate of 60 Hz. This *1125/60 standard* compares with the type of resolution given by projecting a 35 millimeter formatted film onto a large screen and establishes 1035 scanning lines in the active picture display. Also, as an international standard, the 1125/60 systems fits within the need to convert from older systems that have 525 and 625 scanning lines.

Thus, the 1125/60 standard allows existing television signal distributors to convert from the NTSC 525/59.4 standard through readily available large-scale integrated circuits and establishes a format for the global distribution of video information. Currently, the HDTV broadcast system shares television bands with existing services and utilizes unused channels. With this, television signal broadcasters are temporarily assigned a second channel to accomplish the transition from the NTSC format to the HDTV format.

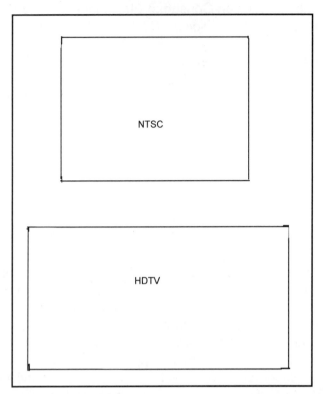

Figure 8.11. Comparing the NTSC and HDTV aspect
ratios.

Other Digital Television Standards

Along with the 16:9 aspect ratio and the 1125/60 scanning refresh standard, the
HDTV design team also determined that the new system should have:

*   2:1 Interlaced scanning combined with non-interlaced scanning

*   A luminance bandwidth of 30 Hz

*   Two color difference signals with bandwidths of 15 MHz

*   An active horizontal picture duration of 29.63 microseconds

- A horizontal blanking duration of 3.77 microseconds, and

- A new sync waveform

The HDTV standard assembled by the Grand Alliance takes advantage of the interlaced scanning used for television transmission and reception and the non-interlaced scanning commonly seen with computer monitors. With non-interlaced, or progressive scanning, the HDTV system provides a choice of 24-, 30-, and 60-frame-per-second scanning with a 1280 x 720 pixel dot resolution and a 24- and 30-frame-per-second scan with a 1920 x 1080 pixel dot resolution. As a whole, HDTV supports the following spatial formats:

1280 x 720:	23.976/24 Hz	Progressive
	29.97/30 Hz	Progressive
	59.94/60 Hz	Progressive
1920 x 1080	23.976/24 Hz	Progressive
	29.97/30 Hz	Progressive
	59.94/60 Hz	Interlaced
	59.94/60 Hz	Interlaced

With this, the HDTV system provides direct compatibility with computing systems. In addition to the non-interlaced scanning formats, the system also offers 60-frame-per-second interlaced scanning at a resolution of 1920 x 1080. The use of interlaced scanning becomes necessary for the two 1920 x 1080 x 60 formats because of the lack of a method for compressing the formats into a 6 MHz channel. Each of the formats features square pixels, a 16:9 aspect ratio, and 4:2:0 chrominance sampling.

Comparing the Broadcast Formats

1080i Interlaced Scanning

As mentioned, *Interlaced scanning* is a process where electron guns draw only half the horizontal lines with each pass. With one pass, the guns draw all odd lines while the next pass draws all even lines. As a result, one complete frame of information is created for every two fields scanned. With field generated every 1/60th of a second, the human eye cannot discern the scanning motion. To compensate for any possible flicker, manufacturers of interlaced scanning displays choose phosphors that have a higher decay time.

Because interlaced scanning refreshes only half the lines at one time, it can display twice as many lines per cycle. Thus, the display technique provides an inexpensive method for yielding more resolution. Interlaced scanning has a relatively slow trace and retrace time that affects the ability of a display to show animations and video graphics.

720p Progressive Scanning

With *progressive scanning*, every line of information on the display is scanned by the electron gun at each pass across the panel. The lines are transmitted from top to bottom. The technique enhances the vertical resolution of the display while allowing the viewer to sit closer to the display and provides an image quality close to the quality seen with 1080 lines of resolution. Viewing distances with progressive scanning shorten to 2.5 times the height of the display. Progressive scanning allows the transmission of 24 frames per second and—as a result—the squeezing of more channels into available bandwidth.

480p Standard Definition

SDTV uses progressive scanning to scan 480 lines one after the other onto the screen. Because of the lower number of horizontal lines, the SDTV format does not require large amounts of bandwidth. The SDTV format allows the transmission of multiple programs in the space traditionally used by one channel or the transmission of data services within the band allocation.

International Digital Television Formats

Along with the different format preferences seen with domestic television broadcast providers, still another difference exists on the international scene. In the United States, the FCC established a rule making committee called the Advisory Committee on Advanced Television Service, or ATSC. In addition, the FCC decided that new HDTV signals would be broadcast on currently unusable channels and that broadcasters would be temporarily assigned a second channel for the transition to HDTV.

The United States, Canada, and Mexico cooperated in the development of the ATSC standard for high definition television and have been quick to adopt the standard for broadcast transmission. In addition, Argentina, South Korea, and Taiwan have adopted the ATSC standard. Other countries considering the adoption of the ATSC standard include most—if not all—of Central and South America, and Russia.

Despite the acceptance of the ATSC standard by those and other nations, other countries have either adopted proprietary standards or have chosen the DVB standard presented by European companies. The Digital Video Broadcasting, or DVB, project began in 1993 with the signing of an agreement by 85 European companies. With this, the companies promote a standard that uses an MPEG-2 transmission scheme. After evaluating both transmission standards, Australia opted for the DVB standard rather than the ASTC standard.

Japan has announced plans to implement its own digital television standard by 2003 while China has test a digital television format that combines the ATSC and DVB standards. In the early 1980's, the Japan Broadcasting Corporation, or NHK, proposed the MUSE HDTV interlaced system that would use 1,125 scan lines and introduced it as a possible world standard. With this proposal, NHK established a goal of high-definition television playing on a wide screen format.

The Benefits of Digital Television

The system achieves many of the improvements in resolution and color reproduction through the decision to establish a 30 MHz luminance bandwidth and two color difference signals with a bandwidth of 15 MHz each. In effect, the decision to use the 30 MHz and 15 MHz bandwidths depended on the decision to use 1125 scanning lines. Because of that decision, the system required a bandwidth of at least 25 MHz. From an overall perspective, the combining of increased horizontal and vertical resolution with wider luminance and chrominance bandwidths yielded a larger number of viewable pixels. Given 1920 horizontal pixels, the HDTV system becomes a platform for several different applications of computer display technologies ranging from computer-aided design and manufacturing to medical imaging.

The HDTV broadband, 20 megabit-per-second digital transmission system enables the convergence of the entertainment, industrial, medical, and educational technologies by using a packetized data transport structure based on the MPEG-2 compression format. Each data packet is 188 bytes long with 4 bytes designated as the header or descriptor and 184 bytes designated as an information payload. With this type of high compression data transportation, the HDTV system can deliver a wide variety of video, audio, voice, data or multimedia services and can interoperate with other delivery or imaging systems.

While the digital transmission system may allow the simultaneous transmission and reception of those services, viewers could select services that would substitute for the normal daily programming. For example, a local PBS station could broadcast HDTV programs such as National Geographic specials or ballets during the evening prime time hours while also broadcasting data services such as weather forecasts or stock market information. The weather and stock information would be apparent to viewers who had requested the service. During school hours, the same station could deliver five simultaneous education programs to participating local schools and homes.

Over-the-air broadcasts of HDTV signals will rely on a 8-VSB vestigial sideband broadcast system while cable transmissions of HDTV signals will use a 16-VSB vestigial sideband system. With this, the system minimizes any potential interference between the HDTV broadcasts and conventional NTSC transmissions. Each of the standards use digital technology to provide a high-data-rate transmission and ensure a broad coverage area. The higher-data-rate transmission for the HDTV cable signals allows the transmission of two full HDTV signals in a single 6 MHz cable channel.

Digital Television and DBS Systems

DBS services have the advantage in becoming the first satellite broadcast services with available HDTV programming. However, the advantage also translates into change for the DBS industry. Current DBS signals broadcast with multiple channels on the same transponder. In addition, the bit rate of each individual channel changes as the content varies with compensation occurring for scenes that have greater activity. ATSC signals broadcast by local stations and rebroadcast on satellite channels feature a much higher bit rate for all programming content.

In comparison to the seven or eight channels currently broadcast from a DBS satellite transponder, the wider bandwidth required for HDTV signals limits the number to only two per transponder. HDTV broadcasts will have a fixed 6 MHz spectrum band and transmit data at a rate of 19 megabits per second. As a result, DBS signal providers such as DIRECTV and DISH have opted to purchase space on additional satellites.

However, the purchase of space on additional sites may place HDTV signal transmissions on one satellite and standard format broadcasts on another satellite. As the HDTV services begin, HDTV broadcasts may occur from alternate satellites rather than primary satellites. As a result, the simultaneous reception of standard format broadcasts and HDTV broadcasts through a DBS system may require two reflectors.

The encoding/decoding format used for ATSC signals utilizes a much more complex scheme than that seen with other digitized formats. Because of this, DBS providers have begun to introduce receivers that feature digital broadcast tuners rather than the analog tuners. The implementation of the new tuners allows the reception of standard compressed television signals as well as the HDTV signals but points towards a new generation of integrated receiver-decoders. Current receiver-decoder combinations will not decode and reproduce the HDTV signals.

Summary

While earlier chapters in this book emphasized C-band systems, chapter eight concentrates on digital satellite signals. The first portion of the chapter defined direct broadcast signals, the equipment needed to receive the signals, and the signal providers. In addition, the chapter took the reader through the installation of a DBS system. While much of the installation seemed familiar because of information presented in previous chapters, the DBS installation process involves a much greater integration of hardware assembly and the receiver programming.

As with all new technologies, it is difficult to gauge the initial impact of high definition television on the consumer market. We do know, however, that the inevitable use of digital signals and high definition television formats will cause change to occur in the DBS industry. Moreover, we also found that the new television standard takes advantage of standards and technologies previously assigned only to computer displays. Because of this convergence, HDTV has many applications including entertainment, healthcare, and the simulcasting of broadcast channels.

CHAPTER

Digitizing Satellite Signals

9

As the capability to transmit not only audio and video signals but also program information and other data via digital satellite signals has grown, broadcasters and providers have had to contend with the need for additional bandwidth and storage requirements. More colors, higher resolution, faster frame rates, and overall higher quality have placed additional pressures on transmission and reception systems. As an example, 24-bit color video at 640 x 480 resolution transmitted at 30 frames per second translates into 26 MB of data per second.

Compressing the audio and video information frees space and allows the transmission of 150 to 200 channels within the 500 MHz bandwidth. In addition, the process reduces the amount of power needed to transmit the signal. Depending on the programming content, an uplink station has the capability to increase or decrease the amount of compression. Additional compression works well for scenes that have less complexity and movement. Decreased compression provides a higher quality picture.

Converting Analog Satellite Signals into Digital Satellite Signals

With digital satellite systems, all the video content originates as an analog signal. Equipment at the uplink site converts the analog signals to digital data. During the conversion process, the coding and decoding of the signals consider the frame rate, spatial resolution, color resolution in response to the need for high quality video images. Three processes—sampling, quantization, and binary notation—occur during the conversion of analog video information into a compressed digital signal and ensure that the reproduction of high quality video images exists. *Figure 9.1* shows a block diagram of the process as the encoded signals arrive at the receiver.

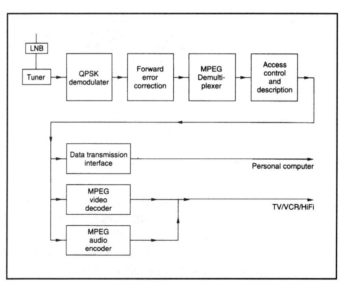

Figure 9.1. Block diagram of the MPEG decoding sequence.

Frame Rates

Every portion of video information breaks down into a series of frames per second. Non-film video production has a standard of 30 frames per second while film video has a standard of 24 frames per second. Each frame splits in half to form a field containing odd lines and a field containing even lines. A standard television displays an analog video signal with the odd field first and the even field second. Sixty interlaced fields show every second. With computer and high-definition television displays, progressive scanning displays each line in sequence and an entire frame 30 times per second.

Color Resolution

As we saw in the last chapter, color resolution refers to the number of colors displayed on the screen at one time. Again, the standard television broadcast has a different color resolution than the resolution seen with computer displays. Designated as YUV, the standard analog video display format displays color resolution in terms of the 7-bit, 4:1:1 or 4:2:2 format that equals 2 million displayable colors and the 8-bit, 4:4:4 format that displays 16 million colors. A 24-bit per pixel computer display shows 16.7 million colors.

Spatial Resolution

Spatial resolution measures the size of the picture. As with color resolution, no direct correlation between analog video resolutions and computer display resolutions exists. As an example, a standard analog video signal displays a full image without the borders seen with computer displays. The NTSC standard used in North American and Japanese television uses a 768 x 484 display while the PAL standard for European television has a slightly larger display at 768 x 576. Because of the differences between analog video and computer video, the conversion of analog video to a digital format must consider each difference to prevent the downsizing of the video image and the loss of resolution.

Sampling Video Data

During the digitization of an analog video signal, sampling of the luminance component occurs 13.5 million times a second or at a rate of 13.5 MHz. With the sampling of the luminance signal, the digitization of the video signal provides an accurate reproduction of any activity found within a monochrome image. Sampling of the chrominance signals occurs 3.375 million times a second or at a rate of 3.375 MHz. The difference in the sampling rates between luminance and chrominance signals occurs because luminance signals contain brightness, contrast, and green signal information. Chrominance signals provide the red and blue color components for the complete image.

As the analog-to-digital conversion process occurs, the equipment takes samples of the signal during a given number of times per second or the sampling rate. The sampling depth represents a given number of bits per sample. With more bits per sample, each sample may have more possible values. For example, a one bit sample depth represents either an on or an off condition while a two bit sample will have an on, off, and two intermediate values. An eight-bit sample depth will have 256 possible values. The minimum sampling depth required for accurate video representation is 24 bits or 3 bytes and provides for a maximum of 16 million colors.

Sampling Audio Data

For audio, each sound frequency found in the sample corresponds to a digital value. The conversion of music into a digital format requires a sampling rate of 44 thousand samples per second and a sampling depth of 16 bits. With each sample equaling two bytes of information and 44 thousand samples per second, every second of music results in 88 kilobytes of digital information. Sampling also applies to the reproduction of a video signal in a digital format. However, rather than generating an audio signal, the task changes to the generation of a raster image that includes varying amounts of brightness, color, and movement.

Quantization of Video Signals

Quantization assigns numeric values to the intensity of the electron beam needed to accurately reproduce a televised image. Typically, the quantization process allocates eight bits for each green, red, and blue image component. In turn, this yields 256 discrete values for each image component. As a result, a string of numbers represents the green signal, another string of numbers represents the red signal, and another string of numbers the blue signal. The three number strings combine to represent the video signal.

Going back to the sampling rates, the green signal includes luminance values and has numbers appearing 13.5 million times a second, with the red and blue signals having numbers appearing 3.375 million times a second each. Each time a number appears it has a value between 0 and 255. All possible combinations of these three values yield 16,777,216 different colors.

Limitations of Sampling and Quantization

Sampling and quantization produce large amounts of digital information. With video signals, each second of video equals 162 million bits of data. Without using some method for compressing the data, even a small portion of a video signal would become too unwieldy for the efficient reproduction of the signal. While the uncompressed video signal data would require huge storage devices, it would also require the use of larger bandwidths to carry even small numbers of channels.

One second of video information contains 30 frames of images while each frame contains 1.2 million pixels at 3 bytes per pixel, or 3.6 MBs per frame. As a result, each minute of video information contains 6.4 GB of data. Within this transmission, the 30 frames of video information per second ensure the quality reproduction of motion and the luminance bandwidth of 5.2 MHz provides the proper reproduction of brightness, contrast, and picture detail. The video signal has 416 horizontal lines of resolution and a signal to noise ratio of 55 decibels. With a horizontal scanning rate of 15.734 kHz, the 5.2 MHz luminance bandwidth places 544 pixels on the active portion of each horizontal scanning line.

Another key measure for the transmission of a video signal is the lines of resolution. As opposed to scanning lines, lines of resolution represent the number of transitions that occur in a space equal to the picture height. With the NTSC system, the transmission of 80 lines of resolution requires a bandwidth of 1 MHz. As a result, the NTSC system with a 4.2 MHz bandwidth has a maximum number of 336 lines of resolution.

The amount of data moving through a transponder depends on the frequency and the type of modulation used for the video and audio signals. Most transponders have a maximum data throughput rating of approximately 500 MB per second. However, the actual throughput may vary because of other information transmitted along with the video signal. The additional information may include interactive program guides or authorization data. Moreover, the transmission of the digital information occurs over an analog carrier.

With the inclusion of the additional information, the available bandwidth for the video signal may drop to between 200 and 300 MB per second. Without compression, a single channel uses approximately 140 MB per second of satellite bandwidth during the transmission of a clean satellite signal. Without compression, the amount of bandwidth needed to carry a clean signal would limit each transponder to two or three channels.

Compressing the Video Signal

During the compression process, different factors affect the amount and quality of the compression. Those factors are:

- Real-time vs. nonreal-time

- Symmetrical vs. asymmetrical

- Compression ratios

- Lossless vs. lossy

- Interframe vs. intraframe

- Bit rate control

Real-Time vs. Nonreal-Time Compression

During compression, some systems capture, compress to disk, decompress and play back video at 30 frames per second with no delays or in real-time. Other systems capture some of the 30 frames per second and play back only a portion of the frames. With nonreal-time compression, insufficient frame rate that occurs at less than 24 frames per second causes the reproduced video to have a jerky appearance. Moreover, the missing frames may contain extremely important lip synchronization data that prevent the correct matching of audio information with video information.

Symmetrical vs. Asymmetrical

Video images compress and decompress either symmetrically or asymmetrically. This refers to how video images are compressed and decompressed. With symmetrical compression, the compression, storage, and decompression of the video images occurs at the same rate. Asymmetrical compression and decompression occurs as a ratio of the number of minutes needed to compress one minute of video.

Compression Ratios

Compression ratios refer to a different quantity than asymmetric ratios. In terms of video compression and decompression, the compression ratio corresponds to a numerical representation of the original video in comparison to the compressed video. As an example, a 200:1 compression ratio represents the original video as 200 and the compressed video as 1. Compression ratios vary with the compression standard. As an example, the MPEG standard usually has compression ratios of 200:1 while the JPEG standard normally has a maximum compression ratio of 20:1 for high quality images. With all standards, more compression results in higher compression ratios but lower video quality.

Lossless vs. Lossy Compression

Two types of video compression exist. As the name implies, lossless compression does not lose any information during the compression process. Decompressing the compressed file produces information identical to the original information. Lossy compression loses some data while compressing a video data stream and relies on the power of the human eye and brain to fill in the lost information. Our eyes and brain allow us to perceive 24 still images shown in rapid sequence as fluid motion.

Interframe vs. Intraframe Compression

The intraframe compression method compresses and stores each video frame as a discrete picture. In comparison, the interframe compression method transmits and receives only the differences between frames and creates a reference frame. During compression, each subsequent frame of the video compares to the previous frame and the next frame. With only the difference transmitted and received, interframe compression substantially reduces the amount of data.

Bit Rate Control

With most forms of digital transmission, the bit rate signifies the speed of transmitted data from transmitter to receiver. Compared to baud rate, the bit rate specifies the number of bits per second carried across a communications channel such as a phone line, serial cable or satellite transponder. The baud rate describes the transmission rate for data within the channel. The digital transmission of signals relies on symbol rates or the sending of two bits at the same time. With this, the baud rate remains the same while the bit rate doubles.

In terms of conserving bandwidth, bit rate control may exist as the most important compression factor. With the transmission of television signals, the quality of the reproduced frames exists as the highest priority. As an example of bit rate control, international standard H.320 videoconferencing COder-DECoders, or CODECs can compress the video plus audio into as little as 56 kilobits per second or about 1.2500[th] of the original signal and provides a low-detail image at about one video frame per second. Business grade videoconferencing typically requires transmission of at least 256 kilobits per second for good quality pictures at about 15 video frames per second. Excellent motion handling near 30 frames per second is possible at transmission rates of 768 kilobits per second and higher.

With DBS satellite broadcast systems, a statistical multiplexer manages the signal transmission of each transponder at a rate of 23 megabits per second. In turn, signal transmission to several video CODECs that comply with the MPEG2 standard. The CODECs convert uncompressed analog audio and video signals to digital signals at about 140 megabits per second and then reduce the actual bit rate required to between 1.5 megabits per second and 15 megabits per second for DBS applications.

DBS Systems and MPEG Compression

Transmitting and broadcasting digital satellite signals requires the digital uplink control center illustrated in *Figure 9.2*. Located in Castle Rock, Colorado, the 55,000 square foot uplink center has 24 hour a day, 7 days a week operations and a system of full uninterruptable power supply that ensure round-the-clock simultaneous access to 200 digital audio and video channels. In terms of hardware, the uplink center utilizes a 512 input x 512 output routing switcher that carries four audio signals with each video signal. As a result, the uplink center produces a 1024 input x 1024 output virtual signal routing matrix. In addition, the uplink center uses more than 400 miles of audio and video cable.

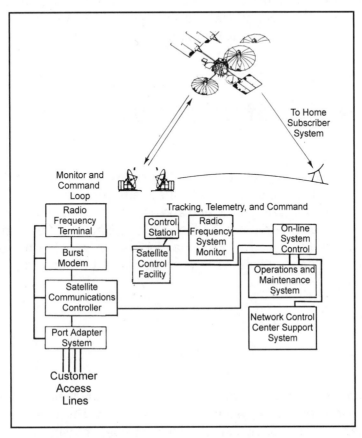

Figure 9.2. Block diagram of the uplink control center.

The fact that multiple channels are broadcast using the same transponder allows yet another level of compression tradeoff to be made. Unlike a single source MPEG compression, a broadcast service can adapt the rate control for several image and audio sources simultaneously. Adaptive rate control spanning multiple programs can give the highest bandwidth to complex sources while reducing the bandwidth to another source, but requires a more complex uplink control facility. However, this type of statistical multiplexing will sometimes lead to blurs or artifacts when the combined ideal bandwidth of the programs on a single transponder is too high. In these cases, one or more programs will have their bit rates dropped to meet the real bandwidth of the transponder. By careful program scheduling, broadcasters can minimize the impact of the fixed transponder bandwidth on the broadcast quality.

As a complicating factor, adaptive rate control of multiple sources requires real time encoding for some of the programs. Thus, compression and quality will not be as high as possible with nonreal-time compression. In addition, broadcasters also encrypt the signal to reduce piracy of programs at the uplink facility. Regardless of these complications, the real world results shown by DBS providers are excellent and closely match what we've been able to produce in the laboratory from the viewer's perception.

The MPEG Standard

The Motion Pictures Experts Group mission grew out of earlier standards work for digital compression of still pictures. In 1988, the international MPEG committee worked towards a goal of standardizing video and audio formats for compact discs and based the MPEG-1 format on progressive sources such as film. By 1990, the MPEG committee had developed a data structure syntax for Source Input Format, or SIF, where video and compact disc audio utilized a combined data rate of 1.5 megabits per second.

MPEG-1

During operation, the MPEG-1 compression system nearly matched the quality seen with VHS video tape playback systems. Even though MPEG-1 worked for film, it did not yield the same results for interlaced broadcast video transmissions. The MPEG-1 standard defines the ability to process fields with a resolution up to 4095 x 4095 and bit rates of 100 megabits per second. In addition, MPEG-1 defines a bit stream syntax for compressed audio and video optimized to not exceed a bandwidth of 1.5 megabits per second. With this, the compression standard has bandwidth restrictions that fit the capabilities of single-speed uncompressed CD-ROM and Digital Audio Tape specifications.

MPEG-2

During 1992, more than 200 international companies contributed to the MPEG draft development that demonstrated strong support for a new technology specification. As opposed to MPEG-1, MPEG-2 includes enhancements that cover the compression of broadcast interlaced television signals. From there, MPEG-2 has become a compression standard for HDTV, DVB, and for high density compact disc technologies. When considering direct broadcast satellite systems, both MPEG-2 and MPEG-2 "near compliant" stand as standard approaches to video and audio signal compression.

MPEG-2 Operation

Figure 9.3 shows a block diagram of the MPEG-2 encoding sequence. During operation, the MPEG compression process examines each frame and compares the contents of the frame on a pixel-by-pixel basis to the previous frame. If the pixel values in the new frame match the pixel values in the original frame, the MPEG compression processor deletes the new frame. In addition, the MPEG processor also examines a range of pixels for areas of identical color and tone.

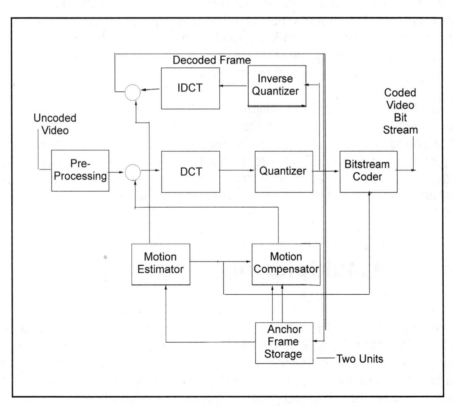

Figure 9.3. Block diagram of the MPEG-2 encoding process.

When two frames match, the MPEG processor inserts a special small marker that instructs the MPEG decompression processor to restore the pixel. If frames have pixels with identical color and tone areas, the MPEG processor removes the duplicate areas and sends on only one pixel. An instruction accompanies the lone pixel and contains information needed to replicate the color and tone areas a specific number of times.

Although the MPEG compression scheme cannot condense volumes of data to a low enough level, the lossy characteristics of MPEG take advantage of the ability of the brain and eyes to fill in any information gaps. As a result, the compression encoder can look for near matches rather than exact matches of data. Adjusting the nearness of the match also controls the amount of compression. However, compression of a file at a ratio higher than 3:1 will result in the averaging of intermediate tones and allows pixelation, or the placement of visible blocks in an image, to occur.

While eliminating visible pixelation, compression ratios limited to a maximum of 3:1 also limit the number of channels per transponder to six. However, most DBS broadcasters use seven to eight channels per transponder through the use of increased compression on video streams that have lower activity. To accomplish this, broadcasters mix the content on each transponder. As an example, a transponder may contain channels that carry game shows as well as action movies with the game shows having much less activity than the movies.

Digital satellite systems not only utilize the MPEG compression scheme but also include encoder circuitry that limits the amount of data. When a large amount of activity in the reproduced picture causes the number of matches and near-matches to decrease, the encoder software responds by lowering the precision of the compression. In addition, digital satellite system compression schemes utilize a statistical multiplexing process that multiplexes channels found on the same frequency together.

Statistical Multiplexing

Statistical multiplexing gathers statistics on the amount of data in each channel and feeds the information back to the MPEG compression system for the purpose of balancing the load. If one channel has low amounts of activity in the reproduced picture and higher match rates, the system allows more precise matches to occur in the paired channel that has more activity and lower match rates. The opposite occurs for channels that have high amounts of activity in the reproduced picture. As a result, the system parameters, selected compression ratios, and the mix of channels affect the compression scheme. Frame-by-frame analysis, compression, and load balancing occur in real time.

MPEG Encoding

During the first step of the encoding process, the encoder circuitry reduces the active area from the NTSC format of 704 x 480 down to 352 x 240. In addition, the circuitry translates the color information from the NTSC to the RGB to the YUV format. While the translation compacts the color signal, the YUV format also separates the color information into independent brightness and hue values. When the human eye views any color, the luminance, or brightness information exists as the dominant component. Because hue is less significant than brightness in terms of color, the MPEG encoder eliminates 75% of the chrominance values.

During the compression process, any unneeded information is immediately discarded. As an example, the NTSC broadcast format uses only 480 out of the possible 525 scan lines to hold image information. The additional 45 scan lines contain information needed for the analog transmission of the signal but unnecessary for the digital transmission of the signal. Discarding the 75% of the chrominance values and the analog-only information allows the MPEG compression scheme to work with only 124 megabits of video data per second rather than the original 162 megabits of information.

With the data, flag, identifier, and error correction information, a data packet for the DBS system is 147 bytes long. The first two bytes of the packet contain the Service Channel Identification, a 12-bit number that ranges from 0 to 4095 that identifies the channel for the packet. The third byte contains the four-bit long flag, an identifier for the encryption code. The next 127 bytes contain the video data. The final seventeen bytes are used for forward error correction.

Discrete Cosine Transformation

The MPEG compression standard relies on discrete cosine transformation, or DCT, for the translation of 8 x 8 blocks of image pixels into sets of numbers. The DCT technique works by removing redundancies from the images. Rather than compress the image, the complex mathematical process of discrete cosine transformation changes the video signal into a form that easily compresses. During the process of compressing the NTSC signal, the video frame divides into 8,100 small individual blocks or boxes. In comparison, the compression of a PAL or HDTV signal results in more blocks. The process also groups the blocks together into five columns called macro blocks and then moves the blocks into an order that increases the efficiency of the compression.

At the beginning of the process, all information about the top left pixel in a given block stores in its complete form. In the next step of the process, only the difference between the next pixel and the base pixel stores. For example, the transformation of a signal that represents only the blue sky, the difference between the first pixel and the next pixel would equal zero. The zero value for the difference would continue for all the pixels in the block and only the difference values would store. If the televised scene includes white and gray clouds, the difference value would change. With the clouds in the lower half of the block, the bottom pixels might differ from the base pixel by changes equaling a negative one or a positive three. After the completion of the process, the values of each pixel in the block are weighted according to the specifications of the system.

Each set of pixels describes one level of detail with low detail images represented by many zero values and high detail images represented by fewer zero values. In essence, each value given through the discrete cosine transformation represents energy at a specific frequency. Rounding off the results of the transformation reduces the number of possible values and produces a better chance for identical values. Although transformation is not the same as compression, frames that feature lower detail compress more than frames with higher levels of detail.

Adaptive Quantization and Variable Length Coding

During the compression process, the system approximates the amount of compression it needed for each block. To do this, the compression system looks for long strings of zeros in the data. Rather than storing the entire string such as 00000000000000, the system stores a phrase that means "15x0." Consequently, the more consecutive zeros contained within the data translates into saved space. Because long strings of zeros rarely occur in practice, the compression code tries to round off all values from negative one to positive one down to zero. If the data stream remains too large, the circuit will attempt to round off all values from positive two to negative two down to zero. This process continues until the proper amount of data compression results in the reproduction of only about 25 megabits per second of video information.

At this point, quantization has a major impact on the size of the final encoded video stream. Larger constants offer fewer possible values and increase the compression ratio. The increased density of compression allows the loss of information and the degradation of video signal quality. Quantization ensures that the video stream data rate never exceeds the throughput of the target output device. During operation, the encoder accesses quantized and transformed values from the lowest to the highest frequencies. As a result, the quantized data becomes strings of identical values that comprise a single token. In turn, each token indicates a value and the number of times that the value repeats.

Encoding the tokens involves the assignment of the most common tokens to symbols that have the shortest possible length. With this process, further compression of the data occurs through the distribution of token frequencies. The MPEG standard further increases the compression by eliminating any redundant data that appears on more than one frame. Throughout the encoding process, the MPEG encoder has the capability to look ahead as many frames as desired to seek repetitive blocks. A pointer that references a single copy of the block replaces identical blocks of pixels common to two or more successive frames.

Forward Error Correction

Encoding a digital signal onto an analog carrier frequency involves the breaking up of data into discrete packets because any transponder frequency will be carrying several different data streams simultaneously. During the encoding process, the multiple streams must interleave in an organized manner so that the receiver can reproduce the data in the same form. Interleaving of data in this fashion is referred to as Multiple Channel Per Carrier, or MCPC.

The use of interleaved audio and video data requires the transmission of additional information along with actual information. A packet identifier, or PID, fits into the data stream and informs the receiver about the type of data. Synchronization packets keep the audio and video information synchronized while system information packets carry other vendor-related information.

Because propagation delays and real-time transmission can cause the dropping out of packets, the compression and decompression of video and audio information relies on error correction. With each transponder handling 40 megabits per second of data, two modes of error correction operation exist. While the high mode uses 30 megabits per second for information and 10 megabits per second for error correction, the low mode uses 23 megabits per second for information and 17 megabits per second for error correction. Because of the rate differences, the high error correction mode requires 3 decibels more of power.

Propagation delays due to atmospheric conditions increase the attenuation of digital signal. Greater path loss causes an increase in the bit error rate of the signal and a reduction of the accuracy of the recovered received signals. In addition, the use of video compression for all the DBS satellite's transmissions translates large numbers of video and audio data bits into fewer numbers of video and audio data bits. Without error correction, propagation delays would cause some of the data to drop out. As a result, the reproduced video would resemble a patchwork of colored blocks or a frozen image.

Forward error correction ensures that DBS can function reliably with small reflectors and without intervention from the transmission site. Without FEC, 18 inch diameter reflectors would otherwise only provide marginal quality receive signals. The generation of the FEC information-signal takes place during the actual transmission of the television signals to the satellites. During this process, a complex mathematical process called a syndrome produces and codes the continuously changing details of the television data signal as a separate part of each television signal.

The syndrome contains sufficient information to allow the decompression processor to correct and fully recover its own original information. At the receiver, a set off processes review the syndrome and recreate any information lost due to an error. However, an extremely high bit-error rate caused by heavy rains or snow can cause the automatic error correction to fail and result in the reproduction of incorrectly placed small picture blocks, blocks of colored snow, or static-like clicking sounds. Generally, a receiver will mute both the sound and video automatically if the error correction fails.

Moving the Data

Because of the large quantities of data handled by digital video equipment, a standard, high-speed interface has become practical for the transfer of digital information. Developed by the Institute of Electrical and Electronic Engineers, the IEEE 1394 standard allows the transmission of digital information between an interconnected VCR, television, computer, DVD player, DBS receiver, computer, printer, camcorder, CD player, and audio amplifier. Most of the electronics manufacturing industry has incorporated the IEEE 1394 standard into their product lines.

The IEEE 1394 High Performance Serial Bus exists as the only standard system that can handle the data volume seen with a video signal. IEEE 1394 transfers 100 to 400 megabits per second and has the potential to transmit at rates of 400 megabits per second and 1.2 gigabits per second. In comparison, the RS-232 interface can transfer data at about 20 kilobits per second while the new Universal Serial Bus can transfer data at 12 megabits per second.

Decoding the Compressed Signal

At the receiver, decoder circuitry restores the deleted information so that the reproduced picture matches the original scene. A digital-to-analog converter translates the video signals from the digital format to the NTSC-standard analog format. Audio signals travel through an MPEG decompression IC and then through another set of digital-to-analog converters for each stereo channel.

The decoder found either in the receiver or a set-top box must recover and maintain specific timing information and perform complex data processing during the decompression and decrypting of the video and audio information. In addition, the decoder processes the digital-to-analog conversions. The decoder also sends pay-per-view billing information back to the signal provider and contains reprogrammable processors that allows the broadcaster to download new code into the decoder automatically.

Compressing the Audio

MPEG-2 AUDIO

The MPEG-2 Audio Standard for low bit rate coding of multi-channel audio supplies up to full bandwidth left, right, center, and two surround sound channels plus an additional low frequency enhancement channel, and up to seven commentary/multilingual channels. Due to the need for compatibility, the MPEG-2 audio compression standard also extends the stereo and mono coding of the older MPEG-1 Audio Standard to half-sampling rates of16 kHz, 22.05 kHz, and 24 kHz for improved quality for bit rates at or below 64 kilobits per second per channel.

Dolby Pro Logic

The Dolby Surround Pro Logic matrix system combines the left, center, right, and a limited bandwidth surround channel into two channels. During playback in a mono-phonic-only system, the two channels go through a summing process. When used in a stereophonic system, the two channels serve as the right and left input channels. Feeding the two channels into a Dolby Pro Logic Decoder allows the matrixed four channels to unfold and become available for playback.

The design of the Dolby Surround Pro Logic system places the matrixed signal within the stereo signal. As a result, the Pro Logic system functions within stereo television broadcasts, the transmission of audio satellite signals for both C-band and digital satellite systems, stereo cable transmissions, stereo FM radio transmissions, stereo laser disc, stereo video tape, and video games. All Dolby Digital AC-3 decoders include digital Dolby Pro Logic decoder circuitry.

Dolby Digital AC-3

Direct Broadcast Satellite systems use the Dolby Digital AC-3 format to transmit 120 stereo music channels to business and commercial establishments through a single transponder. In comparison to the Dolby Pro Logic format, the Dolby AC-3 format takes advantage of Digital Audio Coding, a type of perceptual coding. In practice, perceptual coding seeks to eliminate the data that a human ear cannot hear while maintaining desired data. Digital Audio Coding allows the use of lower data rates with a minimum of perceived degradation of sound quality. As a result, perceptual audio coding places more information into the available spectrum. In addition, the Dolby Digital version of perceptual coding handles multi-channel audio.

While Dolby AC-3 works as a compression format, it also applies superior noise reduction techniques through the lowering of noise when no audio signal is present. With an audio signal, strong audio signals cover the noise at all frequencies through auditory masking. Generally, noise reduction occurs only at nearby frequencies. To accomplish this task, Dolby Digital AC-3 divides the audio spectrum of each channel into narrow frequency bands that correlate closely to the frequency selectivity of human hearing. As a result, coding noise is very sharply filtered and remains close in frequency to the audio signal being coded. The audio signal masks the noise and causes the noise to remain imperceptible to human hearing. With no audio signals present for masking, Dolby Digital AC-3 reduces or eliminates the coding noise.

All this occurs through the use of a "shared bitpool" arrangement where bits distribute among different narrow frequency bands. Dolby Digital AC-3 can process at least 20-bit dynamic range digital audio signals over a frequency range from 20 Hz to 20,000 Hz ±0.5. The bass effects channel covers 20 to 120 Hz ±0.5 dB. In addition, Dolby Digital AC-3 supports sampling rates of 32, 44.1, and 48 kHz. To answer the needs of a wide range of applications, data rates range from as low as 32 kilobits per second for a single monophonic channel to as high as 640 kilobits per second.

The distribution of bits differs according to the needs of the frequency spectrum or dynamic nature of the coded program. Auditory masking ensures the use of a sufficient number of bits to describe the audio signal in each band. In addition, bits distribute among the various channels and allow channels with greater frequency content to demand more data than channels with less frequency content. This type of auditory masking allows the encoder to change frequency selectivity and time resolution so that a sufficient number of bits describe the audio signal in each band. Consequently, Dolby Digital AC-3 can use proportionally more of the transmitted data to represent audio. With the Dolby Digital AC-3 standard, higher sound quality and multi-channel surround sound encode at a lower bit rate than required by just one channel on a compact disk.

The use of a sophisticated masking model and shared bit pool arrangement increase the spectrum efficiency of Dolby Digital AC-3. Rather than use data to carry instructions to the decoders, AC-3 can use proportionally more of the transmitted data to represent audio. As a result, AC-3 has higher sound quality delivered over six discrete channels of sound. Compared to Dolby Surround Pro Logic, AC-3 also includes Left, Center and Right channels across the front of the room and discrete left surround and right surround channels for the rear of the room. All five main channels have a full frequency range of 3 Hz to 20,000 Hz. A subwoofer could be added to each channel, if desired.

A sixth channel, the Low Frequency Effects Channel, sometimes contains additional bass information used to increase the audio impact of scenes such as explosions and crashes. Because the sixth channel has a limited frequency response of 3 Hz to120 Hz, it is sometimes referred to as the "0.1" channel. All six channels in a Dolby Digital system have a digital format that allows the transfer of data from the producer's mixing console to a home playback system without loss.

Digital Video Broadcast

The European Digital Video Broadcast, or DVB, standard relies on MPEG-2 video and audio. In brief, the DVB standard covers the transmission of MPEG-2 signals from satellites, cable systems, and over-the-air antennas. In addition, the DVB standard governs broadcast information but also systems information, program guides, and the scrambling system. Within the standard, DVB-S works for satellite transmissions, DVB-C is the specification for the transmission of DVB/MPEG-2 over cable, and DVB-T is the specification for the transmission of DVB/MPEG-2 signals over terrestrial transmitters.

Summary

Chapter nine summarized the highly technical topics of analog-to-digital/digital-to-analog conversion and digital compression. The chapter moved through a series of definitions as it methodically described the processes needed to encode digital signals onto an analog carrier and transmit the signals to a satellite and back to a satellite receiver. As the processes were defined, you had the opportunity to become familiar with terms and concepts such as sampling, quantization, and forward error correction.

As the chapter progressed, it explained the differences between MPEG-1 and MPEG-2 compression and listed the benefits of MPEG-2 compression when used for the transmission and reception of video signals. The chapter also provided an in-depth explanation of different types of audio signal compression technologies including MPEG-2, Dolby Surround Sound, and Dolby AC-3. The chapter concluded with a brief overview of the Digital Video Broadcast compression and encryption format.

CHAPTER

10

Encryption and Decryption Technologies

In chapters eight and nine we discussed the conversion of an analog television signal into a digital format, the encoding of that digital signal onto an analog carrier, and the decoding of the digital signal at the receiver. The encoding/decoding processes involves the compression and decompression of video and audio information along with other information including pay-per-view billing, configuration data, and scheduling information. Chapter ten builds on the knowledge that you accumulated by reading chapters eight and nine with a detailed look at descrambling technologies.

The encryption of data transforms the data from an intelligible format into an unintelligible format. While the first portion of the chapter describes the different types of encryption formats, the second portion emphasizes decryption technologies and the functions of an integrated receiver decoder or IRD. With this, the chapter also pulls from information presented in chapter seven. Before defining different encryption and decryption technologies, the chapter describes the processing and use of audio and video signals in a television receiver.

Audio and Video Signal Processing

Descramblers apply a variety of techniques and circuitry to the decrypting of encrypted video and audio signals. Those techniques and circuits build off the need for video clamping, sync pulse restoration, and subcarrier regeneration within a color television. While the encryption of video information may remove the video clamping, sync signals, and pulses or move the color burst signal to a different location in the video signal, descramblers use information included in the encrypted and compressed signal to restore the information needed to reproduce the original scene. In some instances, encryption and decryption techniques use combinations of techniques to manipulate the video signal.

Audio Signals

The audio portion of a television broadcast signal is frequency modulated onto a 4.5 MHz carrier. At the television receiver, the sound IF signal is extracted and attenuated before the IF signal travels to the video detector. The extraction of the sound carrier allows the signal to travel into processing circuits designed to handle frequency modulated signals.

Figure 10.1 uses a schematic diagram to show how the sound IF detection, amplification, and limiting circuit precede the sound IF demodulator. Looking at the circuit, the extracted 41.25 MHz sound IF signal follows a path from capacitor C3 and transformer T1 to Q1, the sound IF amplifier. Diode D1 detects the sound carrier and converts the signal to a frequency modulated 4.5 MHz sound IF carrier.

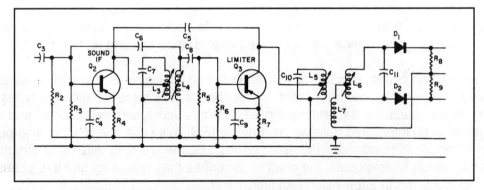

Figure 10-1. Sound IF detection, amplification, and limiting circuit.

After further amplification by the sound IF amplifier, the sound carrier travels into a demodulation stage where an audio detector extracts the audio frequency information from the frequency modulated 4.5 MHz sound IF carrier. At this point, the demodulated audio frequency signal matches the audio signal originally seen at the transmission location. The detector stage accomplishes this transition by developing an ac voltage at its output. Variations in the ac voltage correspond with the frequency variations of the IF carrier found at the detector input.

An *audio frequency*, or *AF, amplifier* is the last stage of the video receiver sound circuit. Unlike the amplifier circuits seen in the past channels, audio amplifiers amplify only the narrow spectrum of audio frequencies that ranges from 20 Hz to 20 kHz. Audio output amplifiers should have the following characteristics:

• High Gain,

• Very little distortion within the audio frequency range,

• High input impedance, and

• Low output impedance

Modern video receivers feature either transistors or integrated circuit amplifiers in the audio output stage that provide enough gain to drive a 30 percent modulated signal. Transistor-based audio amplifiers usually consist of a two-transistor output stage and, in small receivers, have an output range of 100 milliwatt to 1 watt. Audio amplifiers incorporated into an integrated circuit have an output power of approximately two to five watts.

The Composite Video Signal

Figure 10.2 shows the different types of signals associated with the composite video signal. While the sync pulses have a consistent amplitude and spacing, the changing amplitude and spacing of the luminance and chrominance signals represent the changes occurring in a transmitted picture. By definition, luminance signals represent the amount of light intensity given by a televised object, cover the full video-frequency bandwidth of 4 MHz, and provide the maximum horizontal detail.

When we look at a picture reproduced on a television screen, we see an orderly arrangement of light and dark areas. As the electron beam deflects across the inside face of the CRT, the beam action produces a tiny spot of light where it strikes the surface. If the beam stayed at one point, it would burn a permanent spot into the

CRT. However, the beam moves from side to side and from top to bottom as it traces successive, closely spaced horizontal lines. The pattern created by these lines is the raster.

The waveform shown in *Figure 10.2* corresponds with the period where the electron beam traces the last four lines at the bottom of the picture, quickly returns to the top of the screen, and then traces the first lines of the next field. When we break the waveform down into its component parts, the eight lines of picture information are represented at each side of the diagram. At the middle of the waveform, the video waveform is shown as the electron beam returns from the bottom to the top of the picture.

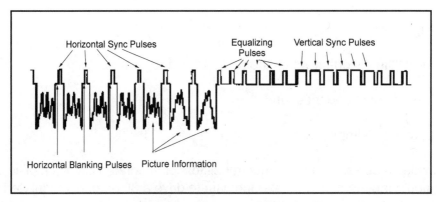

Figure 10.2. Composite video signal.

Luminance Signals

For monochrome receivers, the luminance signal is the video signal. With no need to produce color, the receiver uses amplitude differences of the video signal to produce different shades of gray. Because each color shade has a different amount of light intensity, a multi-colored object reproduced through a monochrome transmission will appear to have different shades. The luminance signal and sync signals separate at the first video amplifier in monochrome receivers.

In a color television receiver, though, the different components of the composite video signal contain the information needed to reproduce both a monochrome picture and a color picture. After the composite video signal leaves the detector, the signal splits into the luminance, chrominance, and sync components. The changing amplitude and spacing of the luminance and chrominance signals represent the changes occurring in the transmitted picture.

Luminance signals consist of proportional units of the red, green, and blue voltages and contain the brightness information for the picture. In addition, luminance signals cover the full video frequency bandwidth of over 4 MHz and provide the maximum horizontal detail. After the weak luminance signal is amplified by the video amplifier stage, the signal travels a relatively short distance to the CRT. Because of the difference in the paths that the luminance and chrominance signals travel, the luminance circuit always contains a delay that allows the signals to arrive simultaneously.

At the video output section, the luminance and chrominance signals add together and proportionately reproduce the complete picture information. When each part of the luminance signal amplitude combines with the chrominance signal, the chrominance signal variations shift to the axis of the luminance signal. To accomplish this, the luminance information is inserted as the average level of the chrominance signal variations. Depending on the receiver, the injection of the luminance signal into the chroma circuits may occur at the three-color demodulators, the video output section, or the CRT. The receiver depicted in *Figure 10.3* has the luminance signal injected at the video output stage.

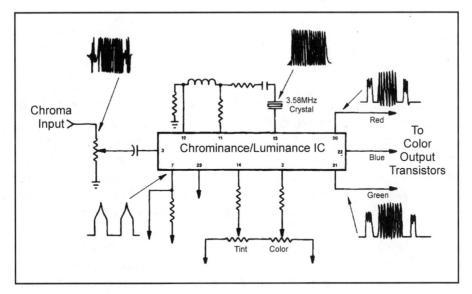

Figure 10.3. Luminance processing circuit.

Processing the Luminance Signal

The composite video signal contains picture, luminance, chrominance, and sync signal information. All the information found within the composite video signal fits within a bandwidth of 4 MHz. This wide band of frequencies is necessary because of the different elements that make up a televised scene.

In any video picture, motion causes a succession of continuously changing voltages to occur. Also, every scene contains different amounts of light and shade that are distributed unevenly across the picture. Each light and shade element corresponds with horizontal lines and vertical fields that, in turn, match with amplitude variations. Because the amplitude variations correspond with the horizontal and vertical scan rates, a horizontal scan contains rapid amplitude changes while vertical scans contain lower frequency variations.

Large areas of a constant white, gray, or black produce signal variations that occur at low frequencies. With no rapid changes in intensity or shade, the amplitude has fewer variations. If we take the same area and break it down into small areas of light and shade, the amplitude changes occur at a higher frequency.

Chrominance Signals

Every color signal represents hue, saturation, and luminance. Essentially, the *hue*, or *tint*, is the color and is represented as an angular measurement. As an example, a blue sweater has a blue hue. When the human eye senses different wavelengths of light for an object, it sees different tints. *Saturation* represents the degree of white in a color and is represented through amplitude measurements. A highly saturated color is intense and vivid while a weak color has little saturation. For example, a red signal that has a low amplitude will have less saturation and will appear as pink.

If we subtract luminance from hue and saturation, we have chrominance. In terms of signals, the chrominance signal is the 3.58 MHz modulated subcarrier signal contained within the composite video signal. Because the human eye cannot detect small color details, the full bandwidth of the chrominance signal ranges from zero to 1.5 MHz and the used bandwidth ranges from zero to only 0.5 MHz. At the transmission point, color information modulates the subcarrier and produces sidebands that extend from 0.5 MHz and 1.5 MHz away from the subcarrier frequency.

A color television receiver contains circuitry that separates the chrominance signals from the remainder of the composite video signal and processes the chrominance signals so that a synchronized color image appears on the CRT screen. The chrominance signals separate from the luminance signals either after the composite video signal exits the detector or after the first video amplifier stage. From there, the chrominance signal travels through a group of circuits that:

- Remove and amplify the color burst signal,

- Reinsert the suppressed color subcarrier,

- Recover the original color difference signals,

- Control the hue and saturation of the reproduced colors, and

- Disable the color signal during a monochrome broadcast

The frequency of the chrominance signal has a mathematical relationship with the horizontal scan frequency so that it also equals an odd multiple of one-half of the horizontal line-scanning frequency.

Frequency Interleaving

When the original standards for the transmission of color television signals evolved, the best method for transmitting those signals was to share the bandwidth with the monochrome signal transmission. However, luminance signals in a color transmission may have a bandwidth anywhere between zero and 4.2 MHz. To provide a transmitted signal with color information, a chroma subcarrier is inserted at 3.58 MHz. Color information modulates the suppressed subcarrier and creates sidebands that extend .5 MHz above and 1.5 MHz below the subcarrier frequency. The sound carrier is located 4.5 MHz from the video carrier and has at least a 0.1 MHz bandwidth.

All this information fits within the 6 MHz channel because of the characteristics of the luminance signal. From 60 Hz to 4 MHz, the luminance signal does not consist of a continuous band of energy. Instead, the signal varies as bursts of energy spaced in 30, 60, and 15,534 Hz intervals. The spacing conforms to the frame scanning rate of 30 Hz, the field-scanning rate of 60 Hz, and the horizontal scanning frequency of 15,750 Hz. In *Figure 10.4*, the frequency components of the luminance signal appear as harmonics clustered around each frequency.

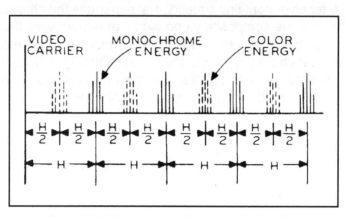

Figure 10.4. Illustration of frequency interleaving.

Frequency interleaving—or the interlacing of odd and even harmonic components of two different signals to minimize interference between the signals—allows the transmission of the chrominance signal within the same 6 MHz channel as the luminance and sound signals. Interleaving of the signals occurs because of the time and phase relationships seen with the luminance and chrominance signals. With the luminance signal, a concentration of energy occurs at multiples of the horizontal scanning frequency. Chrominance information peaks at a frequency of approximately 3.58 MHz.

Luminance signal information appears as an energy burst between every horizontal sync pulse. Chrominance information also appears as an energy burst at the horizontal rate but—because of the introduction of a 3.58 MHz chroma subcarrier—is offset by one-half the horizontal rate. With this offset, the chroma information is displaced by 15 cycles and falls between the luminance signal harmonics.

Referring to the figure, luminance frequency pairs fit above and below the horizontal scanning frequency and are separated by 60 Hz gaps. As a result, a reproduced scene that has 40 pairs of luminance frequencies requires a bandwidth of +-40 x 60 Hz or 2400 Hz. Because of the 60 Hz separation, open spaces exist between each luminance signal cluster. Each of those spaces occurs at an odd multiple of one-half of the horizontal line-scanning frequency. Because of the relationship between the chrominance signal and the horizontal scan frequency, the chrominance signal frequencies fit, or interleave, between the clusters of luminance signal frequencies.

Limiting the bandwidth of the luminance signal to approximately 3 MHz prevents the subcarrier sideband information from mixing with the 4.5 MHz sound carrier and appearing as interference in the luminance portion of the video signal. This would show as patterns in the reproduced picture. Bandwidth limiting prevents the interference from occurring between the low-frequency sidebands and the high-frequency luminance signals. Unfortunately, bandwidth limiting also results in some loss of picture detail.

Video Clamping

The transmitted video signal contains a *brightness reference level* that provides a basis for adjusting the brightness controls of a television receiver. This level may range from the signal level that corresponds with a maximum white raster, the signal level that produces a *black* raster, or a level corresponding with a definite shade of gray. However, almost all transmission standards use the level that produces a black raster as the brightness reference level. The black level remains fixed so that—as the camera moves across a scene—the background brightness varies with respect to black. Every change in the brightness level corresponds with the intensities of the elements that make up the picture.

At the television receiver, the luminance signal consists of an ac voltage component that corresponds with the detail in the reproduced picture and a dc voltage component that corresponds with the average background brightness. Maintaining a picture background that has proper brightness level with respect to black requires that the luminance signal retain the dc component. Using the "black level" as a standard allows any receiver circuit or CRT type to match the varying degrees of brilliance seen with the original picture.

Television broadcast stations use a negative transmission system based on the reduction of power. To make this easier to understand, *Figure 10.5* again enlarges the composite video signal seen at the beginning of the chapter and places the waveform against the horizontal and vertical retrace periods. At the transmission point, the tips of the sync pulses cause the maximum radiated power, blanking pulses cause a 25% reduction in the radiated power, black causes another 5% reduction of radiated power, and white causes the radiated power to decrease to approximately 15% of maximum.

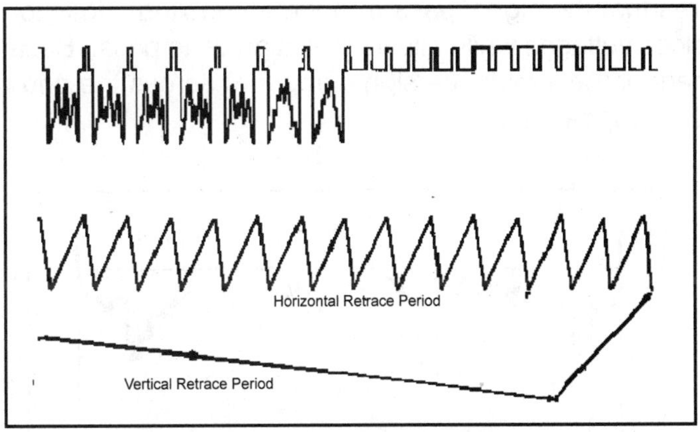

Figure 10.5. Placing the composite video signal against the horizontal and vertical retrace periods.

Every change in the background brightness of a televised scene produces a dc voltage that becomes part of the video signal used to modulate the transmitter. *Figure 10.6* compares the composite video signal for one horizontal line of a dark scene with the composite video signal for one horizontal line of a bright scene. In both figures, the dotted line represents the dc component of the video signal. As you can see, the video signal for the bright scene has a lower dc component than the signal for the dark scene.

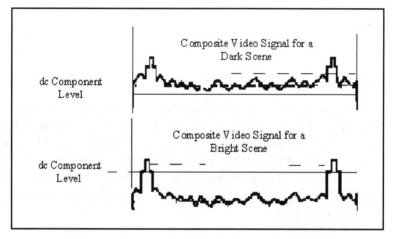

Figure 10.6. Composite video signal for light and dark scenes.

The blanking pulses depend on the polarity of the video signal when cutting off the CRT and producing a black screen during retrace. When a signal voltage with a positive polarity is applied to the cathode of the CRT and the processing circuitry inverts the signal, the negative polarity voltage is impressed on the CRT grid.

However—as *Figure 10.7* shows—inverting the signal voltage and applying it to the grid also requires that the signal pass through a coupling capacitor. As you know, capacitors block dc voltages while allowing ac signals to pass. Because of this, only the ac component of the luminance signal appears across the grid resistor and the picture has a dark appearance.

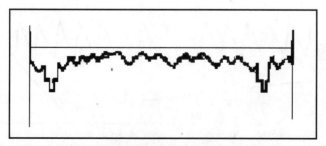

Figure 10.7. Inverted composite video signal waveform.

The blocking of the dc signal component by the capacitor changes the fixed reference for the signal from the blanking level to the average signal level. Consequently, the blanking pulses are less negative than normal and the image details contained within the signal are too bright. In addition, the blanking pulses cannot drive the CRT into cutoff and cause the tube to produce a black screen.

Reproducing the correct image intensity requires that video amplifier stage designs include a method for allowing the dc voltage to control the grid bias of the CRT. Controlling the grid bias allows the blanking pulses to drive the CRT into cutoff. Without the correct dc voltage level at the CRT, several problems will occur because the sync pulses and blanking level no longer have identical voltage levels. The loss of the dc component can cause the reproduction of incorrect colors or an incorrect balance between light and dark scenes.

A clamping circuit inserted between the video output transistor and the CRT restores the dc reference voltage. The *clamping circuit* illustrated in *Figure 10.8* holds the sync tips of the composite video signal to a fixed voltage level. Clamping the video signal to the fixed level restores the signal to the form originally seen at the video detector output.

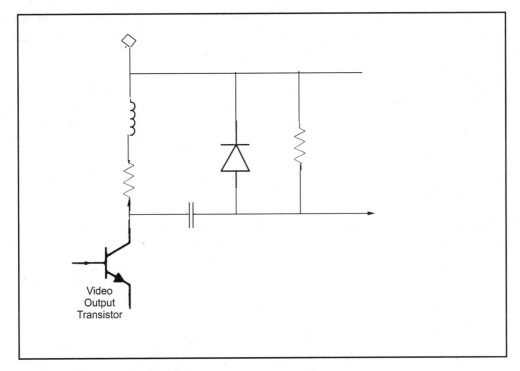

Figure 10.8. Schematic diagram of a clamping circuit.

Sync Restoration

The deflection circuits found in a television receiver rely on separate synchronizing pulses for timing before a raster is produced. The lack of proper timing can cause the reproduced picture to roll vertically or tear horizontally. In many ways, we can consider the sync signals as the structural portion of the picture-making signals. The sync signals include horizontal sync pulses, vertical sync pulses, and equalizing pulses.

Sync separation occurs through the demodulation of the video IF signal, the recovery of the composite video signal, and the amplitude separation of the sync pulses. A *sync separator stage* eliminates the video and blanking signals while amplifying only the horizontal sync, vertical sync, and equalizing pulses. While these circuits separate sync pulses from the picture and blanking signals, sync separation also involves the removal of 60 Hz vertical sync pulses from 15,750 Hz horizontal sync pulses.

Going back to *Figure 10.5,* we can see the complete composite video signal and identify the sync signals. After demodulation occurs, three types of signals—the video, blanking, and synchronizing signals—exist at the output of the video detector. Each of the three signals has a proportional amplitude and exactly the same frequency as the video, blanking, and synchronizing signals found at the transmitter. Applied to deflection circuits in the receiver, the sync signals control the oscillation frequencies of the sawtooth current generators.

The generated vertical and horizontal currents ensure that the electron beam inside of the receiver CRT remains in step with the electron beam within the transmitting camera. Without this structure, the reproduced picture would consist of nothing more than meaningless colored blobs. The horizontal sync pulse rides at the top of each horizontal blanking pulse as shown in the figure and has a duration of 5 microseconds. Spaced 58.5 microseconds apart, the horizontal sync pulses control the side-to-side movement of the electron beam.

The *vertical blanking pulse*—a long duration rectangular wave—carries sync and equalizing pulses on top of the waveform. At the end of each vertical scan, a blanking pulse initiates conditions that reduce the electron beam intensity to zero as beam enters the retrace interval. Once the electron beam scans the last line of picture information, the vertical sync and equalizing pulses occur during the retrace interval.

The *vertical sync pulse* consists of six serrated pulses that occur at the end of each field scan. Within this group of six pulses, each individual pulse has a duration of 27.2 microseconds. 4.5 microseconds separate each pulse. *Equalizing pulses* are another type of sync pulse that occur directly before and after the vertical pulses.

Also grouped as six pulses, the equalizing pulses have a duration of 2.5 microseconds and a 29.2 microsecond separation. Equalizing pulses determine the exact location of the scanning lines of a field in relation to the lines found in the preceding field.

3.58 MHz Chroma Subcarrier Regeneration

Within the *chrominance* components, a 3.58 MHz subcarrier signal modulates the video carrier and produces a side frequency. Any type of video receiver that reproduces color images on a display screen uses demodulation to recover the color and hues from the chroma signal. Demodulation of the color signal involves reinserting the 3.58 MHz color subcarrier signal at the proper phase angles. Reproduction of color and hue is accomplished through a blue minus luminance or B-Y signal and a red minus luminance or R-Y signal. The B-Y signal lies 180 degrees out-of-phase from the burst phase and the R-Y signal lies in quadrature with the burst.

Every possible color reproduced by a video receiver has a phase angle relationship with the 3.58 MHz color burst signal. For example, yellow has a 12.5-degree angle with reference to the zero degree phase angle of the burst signal. Chroma demodulation involves decoding the chroma signal encoded at the transmitter and recovering the specific phasing and amplitude of each chroma signal.

Suppressing the subcarrier leaves the processing circuits without any type of reference signal for detection or for matching the phase of the original transmitted color signal. The burst signal locks the frequency and phase of the 3.58 MHz chroma subcarrier oscillator to the frequency and phase of the oscillator found in the transmitting equipment. With the burst signal in place as a reference, the chroma processing circuits can accurately measure the phase of the chrominance signal. The measurement of the chrominance signal phase translates into the correct reproduction of the hue information.

The color burst consists of 8 to 11 cycles of the subcarrier frequency and rides on the back porch of each horizontal-blanking interval. Although the unmodulated *color burst signal* is included in each horizontal interval, it does not appear on the CRT screen as any type of visible signal. As shown in *Figure 10.9*, the peak-to-peak amplitude of the burst signal equals the amplitude of the sync signal. In terms of timing, the color burst occurs at the same time as the blanking pulse which corresponds with a zero value for the sync pulses. Because of this timing, the color burst signal does not interfere with the sync signals and, as a result, the timing of the deflection oscillators. Later in this chapter, we will examine how the color burst signal becomes a reference for control of the hue.

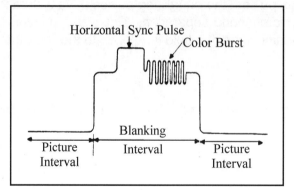

Figure 10.9. Location of the color burst signal.

The 3.58 MHz color subcarrier frequency results from a relationship with the horizontal scanning frequency where the 455th harmonic of one-half of the scanning frequency or:

$$455 \times (15{,}750 \text{ Hz} /2) = 3.58 \text{ MHz}$$

Television receivers rely on 3.58 MHz as a color signal frequency because of several reasons:

- Monochrome receivers use only the luminance signal. With the chrominance signal placed at 3.58 MHz, the high frequency chrominance signal has no effect in monochrome receivers,

- Video receivers rely on heterodyning between the picture and sound carrier frequencies to produce the 4.5 MHz beat frequency,

- Placing the chrominance signal at 3.58 MHz minimizes interfering beat fre quencies produced because heterodyning between the chrominance and sound carrier frequencies, and

- Placing the chrominance signal at 3.58 MHz minimizes interfering beat frequencies caused by heterodyning between the chrominance and medium video frequencies.

Encryption

Satellite programming providers use an electronic method called encryption to control video and audio portions of specific programs. Encryption takes two basic forms called permutation and substitution. With permutation, the order of symbols that represent audio and video signals changes. However, the symbols retain the original meaning or value. Substitution involves the replacement of the original symbols with other symbols. As a result, the new symbols have different meanings or values than the original symbols.

As satellite television technology has evolved, subscription services have grown into a business that covers original programming, movies, sporting and music events, special interest programming, and the news. Encryption allows the programming services to preserve their investment in equipment; new broadcast technologies, and programming by limiting access only to subscribers. To accomplish this, the encryption technologies change the video portion of the signal so that it is not viewable and the audio portion so that it is inaudible. Subscribers to the programming services have access to unaffected video and audio through subscription fees and the use of decoders.

Why Encrypt?

Premium program services purchase the rights to movies from film production companies with the understanding that every individual subscribing to their service will pay for the right to view them. Because of the value associated with movies and other programming, the program providers always have concerns about the theft of the signals. In turn, the application of scrambling technologies also protects local cable and direct-to-the-home operators from the loss of revenue caused through the unauthorized access to their programming. Within a particular region, program producers may license their programs to more than one station. The license agreement may require that the broadcaster limit any access to copyrighted materials through encryption.

Many private business networks also scramble their signals to ensure the confidentiality of information uplinked to and downlinked from a satellite to corporate members attending videoconferences or other non-public meetings. Within the scrambling scheme, corporations may use a series of layered encryption schemes to control the distribution of information within a corporation or among a group of corporations all sharing the same satellite capacity. The confidentiality of information coupled with limited distribution may protect the corporation's status in a competitive marketplace.

Encryption Basics

All encryption or scrambling systems include an encoder, a computerized authorization center, and a network of individual descramblers found at customer homes. An encoder converts standard television signals into separate encrypted video and audio components while an authorization center uses a computer to control all descramblers in the system with each descrambler assigned a unique address code. During operation, the authorization center uses the unique address code to turn individual decoders on or off or to selectively control large groups of decoders. As an example, regional sports networks may use the selective control to black out specific events for regions where the event has not sold out.

The Data Encryption Standard

The data encryption standard, or DES, is a mathematical algorithm residing within a processor and designed for the encryption of data. The DES algorithm relies on a binary number called an active key that consists of 64 binary digits. Of those 64 digits, the algorithm directly uses 56 while 8 digits provide error detection. Encryption codes generated by the processor convert the data into a form called a cipher. While the DES algorithm specifies the enciphering operation, it also specifies the deciphering operation.

When generating the key, the 56 digits used by the algorithm occur randomly while the 8 error correction digits combine with the first 56 digits so that each 8-bit byte of the key will have odd parity or an odd number of ones. Because the 56 digits in the algorithm occur randomly, 70 quadrillion possible keys or combinations of bits exist. Decoding the encrypted data requires the use of the correct enciphering key. Without the use of same key used for enciphering the data, deciphering the data through the use of the algorithm becomes extremely difficult if not impossible.

Conditional Access

Video encryption systems such as those used with direct broadcast systems also feature the use of a smart card. With this approach, the credit card-sized smart card fits into a reader slot located on the chassis front. *Figure 10.10* represents the front panel of a typical decoder and shows the location of the smart card reader slot. The smart card includes a processor that contains mathematical algorithms or deciphering keys. Each set of algorithms allows the descrambler to access subscription and pay-per-view programming. Without the smart card in place, a decoder cannot obtain access to the algorithm needed to address the programming.

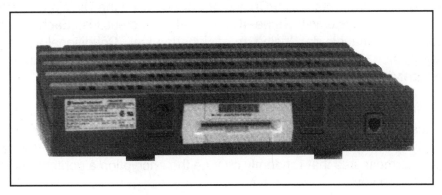

Figure 10.10. Front of a decoder.

Authorizing the receiver involves the sending of an encrypted message to the descrambler that initializes the reception of features or channels. Because the authorization message is a combination of the receiver serial number and the key card number, the smart card will not interchange between receivers. Encryption of the data streams occurs at the uplink site.

When an encrypted packet arrives at the receiver, it passes through the Conditional Access Module before traveling to the demultiplexer. The CAM works as a decryption engine and interfaces with the smart card. During the reception of the encrypted signals, the CAM requests the next set of decryption keys from the smart card. At the beginning of each MPEG-2 packet, a two-bit field called the Transport Scrambling Flags, TSF, provides a value that indicates the presence of either an unencrypted or encrypted data stream.

Of the data streams transmitted to the receiver, some cannot include any type of encryption. Unencrypted data streams provide configuration information for the receiver and include the Systems Information streams such as the Program Association Tables that contain information about each channel and the Network Information Table that points to other transponders used by the service provider. Encrypted data streams include the program guide streams and the MPEG-2 streams.

If the TSF indicates the presence of an encrypted data stream, the MPEG-2 packet passes to the decryption engine. At this point, the CAM uses the key obtained from the smart card to change the packet back into an MPEG-2/DVB transport packet. With the packet conversion in place, the system can process the video and audio information.

Of the major direct broadcast providers, Echostar in the United States and ExpressVu in Canada use the Nagravision conditional access system, while DIRECTV relies on the News Datacomm system. StarChoice of Canada uses the Digicipher system. In practical terms, the systems have nearly identical characteristics. Each encrypts the digitized video and audio data using an enciphering key. Conditional access at the receiver relies on the insertion of smart card containing the correct deciphering key into the receiver and the authorization of the receiver.

Although the conditional access systems have major similarities, differences also exist. Those differences include the size of the key, the mathematical algorithm, and the security of the system. In addition, some services such as DIRECTV utilize electronic countermeasures that randomly change the encryption algorithm and change the length of the key.

Pay-Per-View

The pay-per-view option given through encryption and decryption allows the ordering and delivery of special programming from the uplink center to the home receiver. Most pay-per-view movie services provide movies not released to regular programming services. In addition, pay-per-per-view services provide access to live concerts and sporting events not available on other services. Transactions with pay-per-view services are available through the telephone connection illustrated in chapter eight. The telephone connection serves as the two-way interactive connection between the viewer and the programmer and advises the programmer about the ordering of a pay-per-view event.

Encryption Systems

Scrambling systems used for satellite signals such as the B-MAC, VideoCipher, and Digicipher systems encrypt the video by stripping away the vertical and horizontal sync pulses and inverting the video signal before uplinking the signal the satellite. Other encryption systems also use a Scene Change Detector that alternates the scrambling mode when the video content changes. With this, the encryption system employs multiple inversion modes that invert the video either field-by-field, line-by-line, or randomly. The lack of sync pulses causes the television receiver to lose track of the scanning fields. In addition, some scrambling systems also move the color burst signal to a nonstandard frequency that is detectable only by the descrambler.

Digital encryption converts the analog video information into a digital format. From there, the system cuts and rotates the digitized line segments so that the segments within each line remain shuffled out of order and reassembled at either side of the line segment cut points. Since each line has different cut points, all the vertical information in the picture breaks up in a process that steps the information across the screen in a sequence that changes from field to field. A Pseudo Random Binary Sequence, or PBRS, generator selects the cut points at random and synchronizes the cut points at the uplink transmitter encoder and downlink receiver decoder.

As the uplink encoder periodically interrupts and restarts the final control algorithm, the encoder transmits a special encrypted seed code within the television signal vertical blanking interval to inform the decoder about restarting the algorithm. Synchronizing the restart maintains dynamic locking of the decoder with the uplink PBRS. The encoder generates periodically transmits new seed codes to each authorized decoder in the system. They can be changed at any time to thwart piracy. The decoder performs the complementary cut-and-rotate operation and reconstructs each line at the correct point to reproduce the original picture.

At the descrambler, internal circuitry regenerates the proper horizontal and vertical sync pulses and allows the unencrypted video to display on the television screen. Control of the descrambler and the access to programming occurs through the use of a unique, multi-digit address code that is pre-programmed into the descrambler read-only memory. When the consumer purchases a subscription to a programming service, the address code for the descrambler must accompany the subscription request.

Although satellite television encryption systems scramble the video signal as a way of deterring unauthorized reception, they must also encrypt the audio portion of the television signal. Analog audio encryption systems superimpose the sound information on a low frequency sine wave or invert the audio spectrum so that high frequencies are replaced by low frequencies. In contrast, digital encryption systems scramble the transmission of both the video and audio components of the programming.

With digital encryption, an audio encoder converts the sound portion of each TV program from an analog to a digital signal format at the uplink site. As a result, the audio pitch and intensity are expressed as a burst of pulses, stored and then released over time to recreate the original sound. The horizontal blanking interval of the video signal transmits the digital audio channels and one or more utility data channels.

B-MAC

Designed by Scientific Atlanta, the Multiple Analog Component, Type B, or B-MAC scrambling system stood as a standard for many satellite programming providers until the digital transmission of television signals became more prevalent. The B-MAC system uses a technique called line translation scrambling where scrambler circuitry delays each line of video information by several microseconds. *Figure 10.11* shows the B-MAC signal waveform.

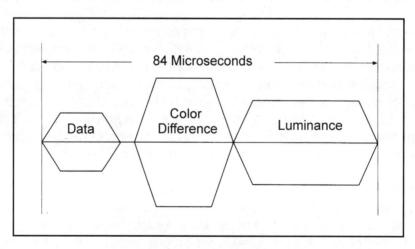

Figure 10.11. B-MAC signal waveform.

In turn, line translation scrambling relies on the time division multiplexing of the luminance and chrominance components of the video signal. B-MAC decoders look for a code addressed by a packet of data contained in the vertical blanking interval of the B-MAC signal before switching on and decoding the signal. The B-MAC horizontal blanking interval contains up to six digitally-encrypted, 204 kilobit per second audio channels and one 63 kilobit per second utility data channel.

The B-MAC system provides up to 256 million addresses with built-in redundancy. B-MAC decoders contain multiple addresses so that the system will serve multiple programmers simultaneously. In addition, the B-MAC system features DES encryption of audio and data.

While the B-MAC encryption scheme provides secure, addressable communications signals, it also improves the signal performance at low carrier-to-noise levels. At the receiving site, the use of the B-MAC scheme results in the sequential transmission of chrominance information in one third the active scanning time of each frame and the sequential transmission of monochrome information in two-thirds of the active scanning time. As a result, the chrominance and luminance signals do not interlace with one another and generate artifacts in the reproduced picture.

Figure 10.12 represents the chrominance and luminance pulses of the B-MAC system. The performance of the chrominance signal improves because only one line of either R-Y or B-Y chrominance information transmits at a time. As a result, the receiver must digitally store the information. To accomplish this, the system filters the chrominance signals and limits the bandwidth to two megahertz. With the R-Y and B-Y components band limited, the system also discards any alternate data samples. As a result, the data rate reduces to 14.31 MHz and the R-Y and B-Y samples store in memory. All this time compresses the chrominance signals to 17.5 microseconds and the luminance signals to approximately 35 microseconds.

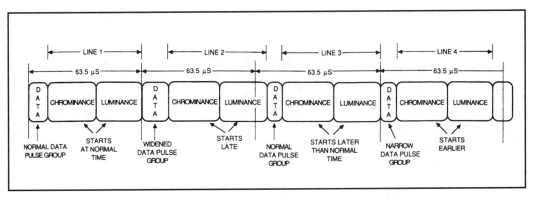

Figure 10.12. Chrominance and luminance pulses of the B-MAC system.

Time Division Multiplex

With time division multiplex, portions of information channels arrive as timed sequenced groups of data. As an example, the sync portion of the video signal precedes the chrominance information that precedes the luminance information. Once the data arrives, circuitry stores and combines the data before converting it into a video signal. The sequential sending of the data also requires the use of time compression. In turn, time compression conserves bandwidth in that time exchanges for bandwidth.

VideoCipher

The VideoCipher I and II encryption formats use a DES algorithm to encrypt and decrypt audio and video signals. With VideoCipher I, the encoder samples active portions of the video scan lines at a rate of 14.318 MHz and then digitizes and stores the samples into memory. Using the DES algorithm, the encoder cuts the individual lines into varying lengths and reads the lines out to a digital-to-analog converter in the scrambled order. The D/A converter converts the scrambled video data back into an analog format. With audio, the VideoCipher I system samples the audio signal at a rate of 44.056 kHz and then uses the DES encoder to encrypt 15 bits per sample. From there, the encrypted audio signal becomes part of the horizontal blanking interval.

The VideoCipher II system builds on the same process used with the VideoCipher I system. However, the VideoCipher II system also features video encryption through the inversion of the video polarity and the moving of the color burst to a non-standard portion of the composite video waveform. Moreover, the VideoCipher II standard inserts an 88-bit data stream and a 3.58 MHz color burst in place of the sync pulse. The 88-bit data stream contains the DES encrypted audio data, program control and billing information, security information, sync data, and auxiliary data that can include teletext, personal messages, dual language capability, and emergency messages. *Figure 10.13* displays the VideoCipher signal.

Because the VideoCipher II system relies on DES encryption for the audio signal and the other information found in the 88-bit data stream, it can provide a large variety of security and system control functions. With each VideoCipher II descrambler containing an individual address, the uplink center sends information that includes the monthly key change along with billing and credit information. The information is encrypted with the descrambler key code so that each descrambler receives individualized information. In addition, individual programs have different keys that indicate program tiers, program ratings, and program costs.

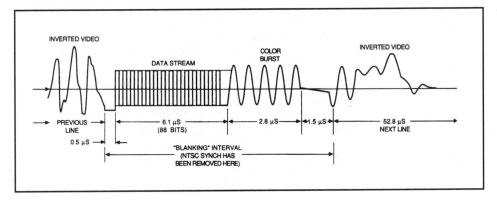

Figure 10.13. VideoCipher waveform.

Using this process, the uplink center can address approximately 600,000 descramblers per hour. The addressing removes the authorization from descramblers for customers who have not paid monthly accounts and authorizes descramblers for new customers. The system also activates and deactivates programming at the request of the subscriber. Each VideoCipher II descrambler can accommodate fifty-six tiers of programming. The program tiers allow the blacking out of rated programs or unauthorized services on an individual or regional basis.

Although the VideoCipher II technology introduced more functionality to encryption and decryption systems, it has quickly become obsolete and replaced by the VideoCipher II+ system. In turn, the new smart-card based VideoCipher IIRS system has begun to overtake the VideoCipher II+ technology. With the VideoCipher II+, coded data that corresponds to the digitized, encrypted audio signal occupies the location once held by the horizontal sync pulse in the VCII system. *Figure 10.14* shows the block diagram for a VideoCipher descrambler.

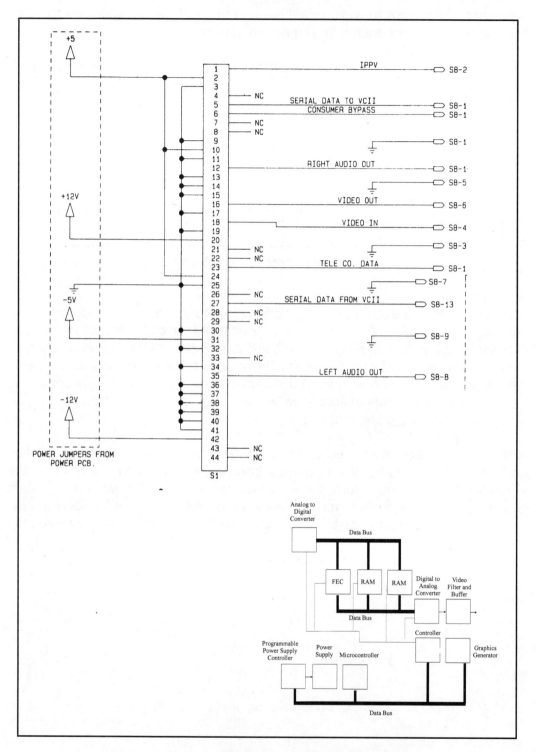

Figure 10.14. Block diagram of a VideoCipher descrambler.

Digicipher

With Digicipher, General Instruments has provided the first commercially available digital compression scheme. The Digicipher format relies on MPEG compression and has become a standard for a number of DBS signal providers. With Digicipher 2, General Instruments uses MPEG-2 coding for video compression and Dolby AC-3 for audio compression. Because of the use of MPEG-2 and Dolby AC-3 for compression, the ASTC digital television format for terrestrial broadcast and used for high-definition television is patterned after Digicipher 2.

The Digicipher system uses compression algorithms that include forward error correction modulation techniques and offset QPSK modulation. As a result, the system can allow up to ten NTSC channels to reside in the space normally occupied by one channel. Digicipher 2 provides true video encryption rather than DES encrypting only the audio as seen with the VideoCipher technologies. In the Multiple Channel Per Carrier, or MCPC, application, Digicipher 2 has a data rate of approximately 27 megabits per second.

The Digicipher 2 system provides CD-quality audio through the use of Dolby AC-3 technology and allows up to four audio channels per video channel. As with the VideoCipher system, Digicipher 2 relies on renewable security cards. In addition, the encryption/decryption system provides 256 bits of "tier" information, has the capability to copy protect and prevent the recording of programming, and allows the insertion of commercial information at the cable company level.

Looking Inside a Descrambler

As you have seen, descramblers decrypt encrypted signals by generating new vertical and horizontal sync pulses, restoring the color burst signal to the correct location, removing the inversion from the video signal, and reconstructing information removed during the compression and encryption processes. *Figure 10.15* provides a block diagram of a sample descrambler and illustrates the process of decrypting the encrypted signal.

At the left of the diagram, the scrambled baseband video input feeds into the input of a video amplifier at J1 and the front end IC depicted in *Figure 10.16*. The circuit contains analog-to-digital converters that quantize the analog baseband video signal at a sampling rate of two times the symbol rate. A control word within the IC sets the initial clock frequency while a QPSK modulator circuit provides clock and carrier recovery. The decoder within the device detects and decodes the compressed MPEG-2 video signal.

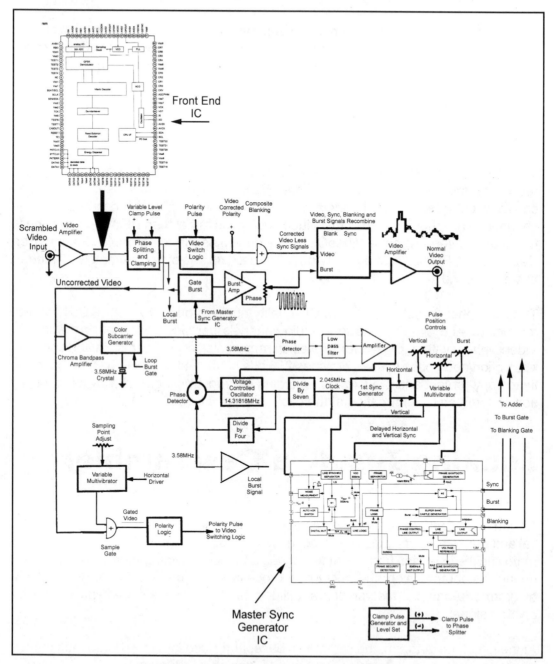

Figure 10.15. Block diagram of a descrambler.

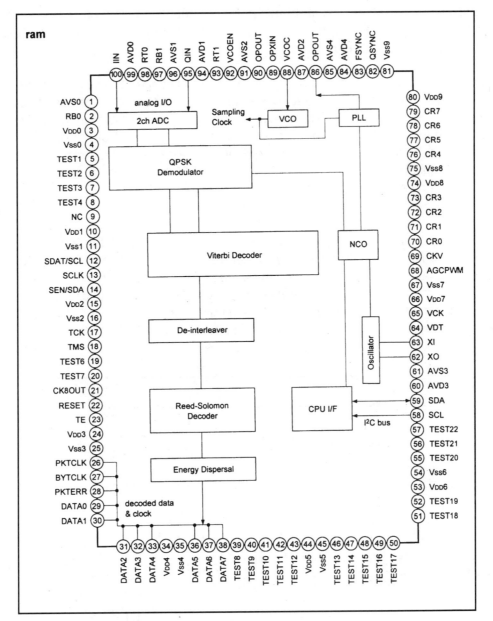

Figure 10.16. Diagram of a front end IC for a descrambler.

After traveling through a phase splitter and clamp, the recovered video signal separates in three directions. Two paths from the phase splitter and clamp take the in-phase and out-of-phase portions of the signal into a two-channel multiplexer circuit. Here, the multiplexer circuit corrects the polarity of the video signal by sensing and automatically correcting any video inversion. At the top right of the diagram, an integrated circuit clamps the blanking pedestal at zero volts while also establishing the dc voltage video level. The in-phase portion of the video signal also feeds into a chroma bandpass amplifier and subcarrier regenerator that amplify the signal and derive the 3.58 MHz reference signal.

At the center of the diagram, the oscillator circuit consisting of the voltage-controlled oscillator, the local oscillator, phase-locked loop, and phase detector generates the burst signal. The phase-locked loop keys are on during the burst intervals and maintain a stable lock on the color burst signal. Moving to the right of the diagram, the phase-locked loop circuits derive the vertical and horizontal frequencies from the color burst signal. During the sync intervals, the 3.58 MHz color burst serves as a timing reference signal for the sync circuitry.

A master sync generator IC such as the circuit shown in *Figure 10.17* generates the burst keying, blanking, and sync signals and has a clock rate tied to the color burst signal. Because the synchronization pulses divide evenly between the black level and the synchronization pulse bottom level, the performance of video signal in noise conditions improves. Other circuitry inserts the newly-generated sync pulses into the scrambled signal after the gated analog switching circuit at the top center of the diagram removes the original, missing, or distorted sync signals from the signal.

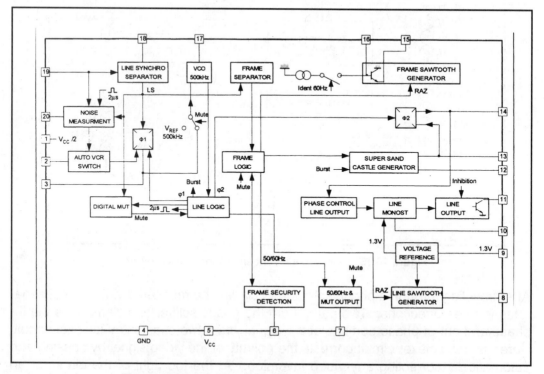

Figure 10.17. Diagram of the descrambler master sync IC.

The frequency divider found at the center of the diagram provides a 7867 Hz reference signal that couples to a frequency multiplier. Within the PLL frequency multiplier, an oscillator runs at 504 kHz. Then, the 504 KHz signal feeds into the master sync generator and provides the necessary timing signals.

At the bottom left of the diagram, a sampling circuit consisting of a gate and multivibrator uses a pulse from the master sync generator to sample the composite scrambled video signal at a reference point. During this operation, the circuitry samples the dc level of the video waveform and feeds the sample into polarity logic circuit where a comparator produces the necessary logic level.

Summary

Chapter ten combined information found through the remainder of book with other material that described the processing of the audio and video signals in a color television to establish the basis for a discussion about the encryption and decryption of satellite signals. The text defined encryption and decryption before illustrating why the encryption of satellite-transmitted signals has become a necessity. Then, the discussion moved to different encryption formats commonly used in the satellite television industry.

Before moving to the detailed analysis of the B-MAC, VideoCipher, and DigiCipher encryption and decryption systems, the chapter provided an overview of the DES, or digital encryption system standards, conditional access, and pay-per-view. The analysis of the encryption and decryption systems moved from the conversion and coding of the original television signals into different formats and the decoding of the signal at the descrambler. The chapter concluded by using a block diagram of a sample descrambler and schematic diagrams of several integrated circuits to illustrate the processing and decryption of an encrypted signal.

CHAPTER

11

Expanding Your System

Chapter eleven introduces several methods for expanding and enhancing a satellite system. The chapter begins by showing how to plan for adding more receivers to an existing system. After stepping through the planning process, the chapter also describes practical methods for installing the receivers. The description covers systems with two receivers, more than two receivers, and satellite master antenna television systems.

Because many readers wish to install DBS systems on recreational vehicles or use portable DBS systems when camping, the chapter provides a complete overview of RV installation techniques. The overview considers available antenna mounts and provides instructions for installing a DBS on a motor home. In addition, the chapter describes accessories that make both the RV and the portable installation easier. The chapter also shows how to use a DBS system as an access point to the Internet.

Before moving to the installation and set-up of the DirecPC system, the chapter defines networking terminology and compares different networking systems and transmission media. The description of computer networks leads to a discussion about how the DirecPC system operates. As the chapter closes, it shows how to install the DirecPC system for individual use and as a proxy server for a network.

Adding Another Receiver to Your System

Many families and businesses add a second receiver to an installation as a way of providing independent viewing of satellite programming from a different location. Whether building or remodeling, the availability of multi-receiver systems has made pre-wiring the home for the reception of off-air terrestrial, satellite, and cable signals a reasonable option. As a result, it has become efficient and effective for homeowners to plan for the installation of cabling and wall jacks used for the reception of television signals in the same way that they would plan for the distribution of electricity.

Planning for a television signal distribution system should consider:

- the use of a single or multi-receiver system

- the possible number of receivers

- the possible location of the satellite dish

- the possible use and location of a traditional off-air antenna

- the possible location for a wiring hub

- the need for any amplifiers or switches

- the plan for routing the cable

- the location wall plates, phone jacks, and electrical outlets

Figure 11.1 depicts a C-band dual feed system that connects each receiver to a separate LNB attached to the feed. As you may recall, odd and even transponders on a satellite alternate from vertical to horizontal polarization as a method for eliminating interference between the signal frequencies. Because of this, the channels received at the home also alternate from vertical to horizontal polarization.

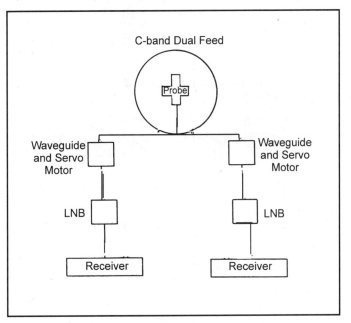

Figure 11.1. C-Band dual feed system.

The receivers connect to separate LNBs and wave guides because of the desire for independent selection of channels requires independent polarization. Inside the feed, a special probe responds to movement from either of the servo motors. However, the use of a dual LNB system also requires the use of a dual LNB mount at the feedhorn and the installation of two RG-6 cables.

Figure 11.2 shows the same type of arrangement with a DBS system. However, the DBS system does not rely on the mechanical movement of a probe in the feed. Instead, voltage differences at the LNB cause the switching from one polarization to the other. When setting the receiver to an odd channel, the receiver supplies the LNB with a lower dc voltage. Switching to an even channel causes the receiver to supply the LNB with a higher dc voltage.

With the two C-band receivers and only one LNB shown in *Figure 11.3*, one receiver would act as a master while the other would act as a slave. That is, the master receiver would supply the voltages to the LNB and receive both odd and even channels because of the ability to switch polarities. The slave receiver would not have the ability to switch polarities and would work with only the polarity used by the master receiver.

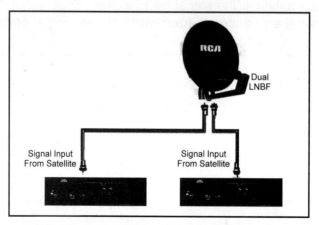

Figure 11.2. Dual LNBF system with DBS receivers.

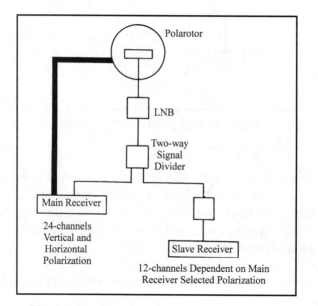

Figure 11.3. Diagram of a master-slave receiver installation.

Moving back to *Figure 11.1*, the connection of each receiver to a separate LNB allows both receivers to receive all vertically and horizontally polarized channels on an independent basis. However, the use of a dual LNB system also requires the use of a dual LNB mount at the feedhorn and the installation of two RG-6 cables. *Figure 11.3* shows a dual LNB mount.

Adding More Receivers to Your System

Another solution for the independent switching of polarities has evolved from the need to connect more than two independent receivers to a system. As opposed to the dual LNB mount, no triple or quadruple LNB feed mounts exist. Instead, install-ers rely on the multi-switch arrangement shown in *Figure 11.4*. As the figure shows, the multi-switch arrangement continues to rely on the use of dual LNBs. The multiswitch automatically sorts the voltages supplied by individual receivers and switches the receiver from one LNB to the other.

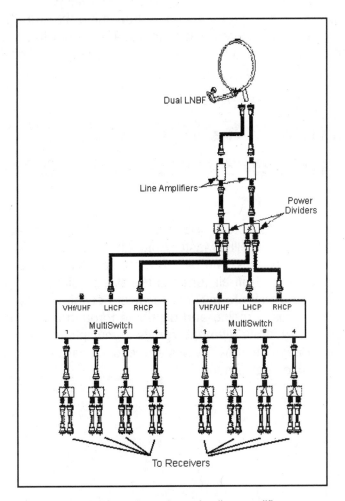

Figure 11.4. DBS configuration using line amplifiers, power dividers, and multiswitches arrangement.

As many as sixteen receivers could attach to a system through a multiswitch. Any receivers switched to an even channel and supplying the same voltages switch to one LNB while any receivers switched to an odd channel switch to the other LNB. The output from the LNB splits between the receivers. *Figure 11.5* shows a large-scale multiswitch arrangement.

Infrared Receiver/Emitters

An infrared receiver/emitter system allows the use of one infrared remote throughout an entire location for receivers that use the same set of infrared frequencies. The emitter portion of the system mounts in cabinets, on shelves, or into light switches and sends the signal from the remote control transmitter back to the source. Individual emitters either connect directly into a block amplifier or function by having the signals inserted into the coaxial cable used by the system and recovered near the equipment.

Installing a Satellite Master Antenna Television System

Businesses such as hotels, motels, and bars rely on another derivative of the multi-receiver system called the Satellite Master Antenna Television, or SMATV, system. A SMATV system features private ownership and the capability to receive satellite television, cable television, and off-air terrestrial television signals and then independently distribute those signals to multiple receivers or televisions. Listed in *Table 11.1*, the different signals combine and distribute through a cabling network. Because of the need to distribute different signals to multiple locations with little or no signal loss, the installation of a SMATV system requires careful planning and the use of specialized distribution equipment.

Type of Signal	Frequency Band
FM Radio Band	90 to120 MHz
VHF TV Band	54 to 216 MHz
Superband Cable TV (Channels 23 to 36)	216 to 300 MHz
Hyperband Cable TV (Channels 37 to 62)	300 to 456 MHz
UHF TV Band & Cable (Channels 63 to 140)	470 to 890 MHz
Satellite Signal Downconversion	950 to 1450 MHz

Table 11.1. Signals at a SMATV system.

The SMATV Head End

Every SMATV system divides into a head end and a distribution system. In turn, the requirements of the system will determine the type of equipment that is needed for the head end. System requirements and design depends on the dimensions of the building and the signal requirements of the installation. A head end consists of the antenna, signal processing equipment, and any required amplifiers. *Figure 11.5* provides a diagram of a sample SMATV head end.

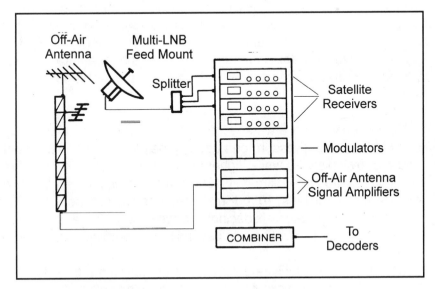

Figure 11.5. Sample head-end.

At the head end, the need for the clear reception of signals requires individual channel control for filtering, setting video and aural levels, and combining. A well-designed head-end reduces cross modulation distortion; eliminates adjacent and co-channel interference; and produces balanced signals ready for distribution. The distribution system determines the requirements of the head end. That is, the adding of all the losses that occur throughout the distribution system indicates the amount of signal that the head end should produce.

The location of the head end depends on the distance to the receiving antennas. Longer distances between the head end and the receivers create a greater potential for signal loss. In addition, a head end should be placed in a temperature-controlled environment that provides accessibility as well as available power.

As shown in an earlier chapter, the distance of the RG-6 cable from the head end to the receiver should never exceed 100 feet. For installations where the cable run distance exceeds 100 feet, amplifiers can boost the signal level. When installing the system, always place the amplifiers in accessible locations.

SMATV Distribution Systems

A well-designed and installed distribution system provides acceptable signal levels to every receiver in the system as well as isolation between the receivers. The distribution system includes the cabling, signal splitters, line amplifiers, line taps, signal combiners, modulators, compensators, and wall jacks. As an analogy, the SMATV distribution system resembles a water distribution system in that pressure—in the form of voltage or decibels millivolts—must remain consistent and stable for the system to function as designed.

Just as insufficient pressure in a water system will allow water to only trickle out of some outlets and too much pressure can cause pipes to rupture, insufficient signal levels will cause snowy pictures while higher than normal levels will overdrive amplifiers. A television requires approximately 1000dBmV of signal to reproduce a snow-free picture. In terms of decibels, 1000dBMV references to zero decibels. For every doubling or halving of the signal, a six decibel change occurs. To regulate the amount of voltage pressure in the distribution, we use splitters, combiners, and line taps.

In *Figure 11.6*, each outlet receives a dedicated output signal from the head end equipment. Along with providing excellent quality signal distribution, the system shown in *Figure 11.6* also offers flexibility for expansion. The system expands simply by adding and connecting additional receivers.

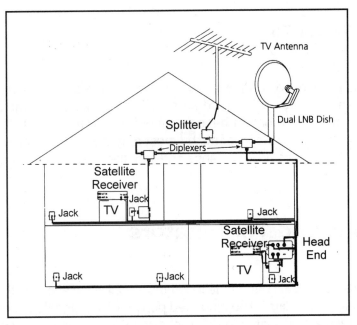

Figure 11.6. SMATV distribution system.

Channel Combiners

The use of a channel combiner allows the simultaneous viewing of channels from different sources. Channel combiners eliminate the need for the manual switching between off-air transmissions, VCRs, cable television signals, and satellite signals by combining the channel 3 or 4 RF output of the satellite receiver with the antenna lead from a conventional antenna or cable. In addition to combining the signal, channel combiners also prevent the reflection of the satellite signal back up the antenna.

Modulators

An external modulator allows the selection of any UHF channel and most cable channels as an output channel for the satellite receiver. As we found earlier, satellite receivers normally output to VHF channels 3 or 4. In addition to providing a larger selection of output channels, an external modulator may also support multiple input signals and the placement of those signals onto different channels. The output signal from external modulators can feed into a channel combiner for more amplification and then feed to different television receivers. Modulators are available as monaural analog or digital stereo units.

Compensators

The distribution of television signals over long cable runs can result in the attenuation of high frequency signals. Displayed as a difference in signal quality between low and high channels, the problem becomes more apparent through the emphasized amplification of low-frequency signals by most amplifiers. Compensators attenuate lower frequencies and provide a balance between the low and high frequencies found in the television signal.

Line Taps and Terminators

Line taps extract a specific part of the incoming signal while allowing the remainder to pass to the main output line. The extracted portion of the signal passes onto another output port. Taps are available in different sizes with the value of the tap used in a distribution system determined by the amount of signal present at the insertion point and the amount of signal needed at the tap. Every tap has insertion loss with the lower tap levels producing higher insertion loss levels.

Terminators consist of an encapsulated resistor fitted onto an F-connector and fasten onto unused ports in a distribution system. The resistor matches the characteristic impedance of thecoaxial cable and prevents any interference from leaking into the system. Non-matched devicescreate a condition called standing waves that reduces the efficiency of the system and causes ghoststo appear in the reproduced picture. With a receiver connected to every output in the system, alloutputs are terminated. Any line not connected to a receiver should connect to a terminator to ensure the proper impedance match throughout the system.

Multi-Receiver Systems and Signal Quality

Splitting and switching signals may or may not have an effect on the signal quality. As opposed to analog satellite receivers that depend on signal level, digital satellite signals depend on precise signals. Line amplifiers provide an obvious solution to the loss of signal in distribution systems. However, line amplifiers also amplify and introduce noise as well as amplifying the signal.

Splitters

Splitters allow the division of the signal from one satellite receiver for multiple television locations. Available in 2-way, 3-way, and 4-way options, splitters are available for either the 900-1500 MHz frequency range or the 5-900 MHz frequency range. The 900-1500 MHz frequency range splitter divides the signal from the LNB and passes the split signal to multiple receivers. 5-900 MHz frequency splitters work with the UHF/VHF/FM frequency bands. Despite the ease of use seen with splitters, signal loss occurs with each connection and adds to the two decibels of signal loss that occurs per every 100 feet of the transmission cable. *Table 11.2* compares the loss figures.

Type	Average Loss
2 way splitter	~ 4 to 6 dB
4 way splitter	~ 4 to 9 dB

Table 11.2. Splitter signal loss comparison.

Line Amplifiers

Line amplifiers can boost a signal by 20 decibels when used for cable runs that exceed 150 feet. A line amplifier uses the same dc voltage supplied to the LNB and does not require an external power supply. Generally, a line amplifier such as that shown in *Figure 11.9* provides the best results when installed at the middle of the cable run. *Table 11.3* compares the amplification given by a line amplifier for a receiver connected through cable run of approximately 150 feet with the signal level at a receiver connected through a cable run of less than 100 feet.

Transponder Number	No Amplification	With Line Amplification
1	92%	92%
2	78%	81%
3	72%	76%
4	86%	86%
5	89%	89%
6	85%	82%
7	83%	83%
8	84%	80%
9	77%	75%
10	89%	86%
11	91%	87%
12	90%	83%
13	82%	80%
14	80%	76%
15	75%	75%
16	82%	85%
17	88%	85%
18	83%	83%
19	68%	78%
20	76%	83%
21	78%	78%

Table 11.3. Line amplification vs. no amplification.

For most transponders, the amplifier has no additional effect and—in some instances—actually decreases the signal slightly. However, the signal levels for channels 2, 3, 19 and 20 benefit from the amplifier.

Connecting an Off-Air Antenna to the System

An Off-Air Terrestrial Antenna receives the broadcast signal from local stations. While most installations rely on one broad band antenna that can receive all VHF and UHF channels, some may use several single channel antennas that point towards stations that lie in different directions. The selection of an off-air antenna depends on the possible number of received channels to be received, the direction and distance from the home or business to the station, whether the signal arrives as a UHF or VHF signal, and the signal strength.

Most multiswitches support the connection of an off-air antenna to the system. In addition, diplexors split and combine signals so that standard broadcast and satellite signals can travel over the same coaxial cable. With a multiswitch, the antenna connection attaches to the designated port on the switch. If such a switch is used, you simply attach the antenna lead to the designated port on the switch. A diplexor inserts into the line before each receiver, separates out the VHF/UHF and satellite signals again, and provides a separate coaxial lead for each signal source.

RF, Composite Video, and S-Video Signal Outputs

Most, if not all, satellite receivers support three different interconnections of S-Video, Composite Video and RF signals. In terms of distribution, each signal type offers distinct advantages and disadvantages. Because RF signals deliver the satellite channel over VHF channels 3 or 4 at the receiver, RF signals stand as the easiest distribution option. However, RF signals have a monaural audio output and the lowest picture quality of the three options.

The use of composite video through RCA phono-type connectors provides much better picture quality and a stereo audio output. However, the use of composite video requires a television receiver that supports the input of composite video signals. The distribution of composite video signals to more than one television becomes more difficult because of the need for audio/video amplifiers. Wireless distribution systems solve some of the difficulty by using radio waves to transmit the audio/video signals to the receivers.

S-video signals provide the highest quality picture through the separation of the red, blue and the green/black video components. However, the distribution of S-video signals stands as the most difficult because the signal components can lose synchronization over long cable runs. The loss of synchronization and signal strength results in red or blue ghosts around any displayed objects.

Installing a Satellite System on a Recreational Vehicle

The size of DBS systems has made the attachment of satellite antennas to recreational vehicles a great option. DBS antenna systems that mount to an RV arrive in all price ranges and range from clamp mounted systems to more expensive mounts. The antennas installed on RVs feature a folding feed support that lowers the LNBF during traveling. The use of a hand crank unfolds the antenna and feed. A roof mount is shown in *Figure 11.7*. A ladder mount is shown in *Figure 11.8*.

Figure 11.7. RV DBS mount for the roof.
Courtesy Winegard, Inc.

Installation of the antenna and feed occurs through the drilling of a one-inch hole in the roof of the RV, cutting the control shaft for the thickness of the roof, and the fastening of the unit to the roof with screws. Despite the comparative ease of installation, the installation requires planning because of the drilling of a hole through the roof. Always find a location that allows the antenna and feed to clear any RV accessories such as an air conditioner or hatches that also reside on the roof. Moreover, the placement of the antenna should allow the placement of the receiver for easy access by the viewers and close to an electrical outlet.

Many motor home manufacturers use a plywood/aluminum sandwich for the roof. However, some manufacturers may apply a top layer of rubber that requires the use of an adapter plate for the antenna mount. Other RVs have fiberglass stretched between supports and require both an adapter plate as well as extreme care. Never walk on the roof of a motor home before verifying that the roof will support weight. When drilling the hole, ensure that the drilling does not cut any of the support beams or wires that run through the roof of the motor home.

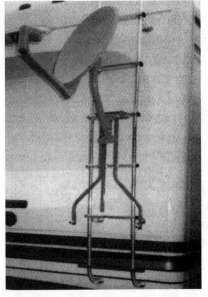

Figure 11.8. A ladder mount for RV DBS installation.

With a few exceptions, elevation, azimuth, and orientation to true North for the antenna occur in the same way as seen for roof-mounted systems. The control mechanism for manual turn satellite systems mounts on the ceiling of the RV and consists of a crank that adjusts the elevation of the antenna, a knob that adjusts the antenna azimuth, and an azimuth indicator plate. After determining the azimuth and elevation for the dish through the installation menu, the viewer raises the antenna from the storage position by turning the crank and observing the signal level meter on the receiver.

Automatic Tracking Systems for the RV

Manufacturers also offer fully automatic RV satellite systems. After reaching the destination, the viewer can simply press a button and cause the antenna to move from the storage position. Entering the nearest city along with the current elevation, latitude, and longitude at the positioner control causes the antenna to adjust for the strongest signal.

Portable Systems

DBS systems are also available as portable systems that pack into a carrying case and attach to the side of an RV with a suction cup. As a variation, manufacturers also offer portable systems that include a flat mount. Along with RVs, the portable systems also work nicely for camping trips and picnics.

Portable System Accesories

Along with offering portable systems, manufacturers also offer a range of accessories for the systems. Power inverters convert the dc voltage obtained from a car cigarette lighter outlet to the ac voltage needed by the receiver. Manufacturers offer the dc to ac power inverters in 300 watt or 1000 watt continuous versions. In addition to power inverters, multi-socket plug-ins attach to the cigarette lighter outlet and provide three 12VDC outlets.

Connecting to the Internet with a Satellite System

Because the DBS signals travel as digital packets, the signal providers can send video, audio, and computer data in any combination from the uplink center to IRDs. The large amount of bandwidth found with DBS transponders also allows DBS companies the option of providing data services such as Internet or interactive TV services. However the greatest drawback occurs because of the asynchronous nature of DBS systems. Currently, DBS broadcasts occur only from the uplink center to the receiver and do not allow the sending of data from the receiver at the same speeds. Any return data from a DBS system travels through the attached telephone line.

The option to display the Internet or an interactive shopping channel on a television requires separate decoders or set-top boxes. Yet, the use of special decoders provides several interesting options for viewers. For example, a decoder with built-in personal television features allows subscribers to set the decoder so that it watches for certain types of favorite channels. Storage devices within the decoder allow the saving of the favorite channels or Internet-based information for later viewing.

Comparing Data Distribution Technologies

The advent of digital communications has allowed the transmission of voice, video, and data information over a single data link. Individuals and companies can communicate through the range of options that include baseband and broadband services such as analog telephone services, local-area networks, wide-area networks, and satellite communications. Communications may occur through copper lines, ISDN connections, T1 and T3 connections, fiber optics, or wireless services.

Broadband and Baseband Systems

The transmission of data occurs in either a broadband or baseband format. With broadband transmission, a single wire can carry several channels at once and uses a broad number of carrier frequencies. Examples of broadband systems include cable television systems, telephone systems, and satellite communications. The baseband format allows the transmission of only one signal at a time. Most communications between computers—including most local-area networks—involves the use of baseband communications and have a transmission distance limited to one mile.

Channels and Data Transfer Rates

With data communications, the term "channel" describes a communications path between two computers or devices. In this general sense, a channel may refer to wiring or to a set of properties that separates one transmission band from another. For example, the use of television channels provides a method for separating frequency bands from another. Data transfer rates describe the transmission speed of data from one device to another. *Table 11.4* illustrates data transfer rate measurements.

Measurement	Abbreviation	Quantity vs. Time
kilobytes per second	Kbps	thousands of bytes per second
megabits per second	Mbps	millions of bits per second
megabytes per second	MBps	millions of bytes per second
gigabits per second	Gbps	trillions of bits per second

Table 11.4. Data transfer rate measurements.

Transmission Media

The transmission media used to carry voice, video, and data communications ranges from the copper telephone lines and the coaxial cable used television signals to fiber optics and wireless systems. Each type of transmission media varies in terms of capacity and carrying distance.

The twisted pair cables used for telephone and local-area network connections have limited bandwidth and work for the baseband transmission of signals. One of the most popular types of transmission media is called unshielded twisted pair, or UTP, and is very similar to the cabling used for telephone connections. The twisted pair portion of the name refers to the use of two insulated wires wrapped around each other. Unshielded twisted pair cabling offers the benefits of low cost, easy installation, and the capability for relatively high data rates.

The standard coaxial cable with a single conductor is also called unbalanced transmission media. Balanced media uses two similar wires to carry signals that have an opposite polarity and are less susceptible to noise and signal distortion.

A T-1 line is a dedicated phone connection supporting data rates of 1.544 Mbits/second. In addition, a T-1 line actually consists of 24 individual channels, each of which supports 64 kbits/second. Each 64 kbits/second channel can be configured to carry voice or data traffic. Most telephone companies allow you to buy just some of these individual channels, known as fractional T-1 access. A T-3 line consists of 672 individual channels with each line supporting a data transfer rate of 64 kbits/second. Internet service providers, hospitals, and corporations use T-3 lines as a main network line and as a connection to the Internet.

Many building-to-building and wide-area network installations involve the use of fiber optics. With this, an insulator encloses a bundle of glass threads. Each thread can transmit information at almost the speed of light. Along the advantage of speed, fiber optics also offers benefits like:

- Greater bandwidth or the ability to carry more data,

- Less susceptibility to interference,

- The ability to carry digital signals, and

- Reduced size and weight

Until recently, the use of fiber optics had not become widespread due to the installation cost and the fragile characteristics of the media. Despite those factors, fiber optics have become more popular for local-area networks and telephone networks. Nearly every telephone company in the nation has either replaced or plans to replace existing copper lines with fiber optics. *Table 11.5* displays the differences between the media types.

Transmission Media	Data Rate	Cable Length
UTP	1-10 Mbps	0.1 km
STP	16 Mbps	0.3 km
Coaxial Cable	70 Mbps	Greater than 1 km
Fiber Optics	100 Mbps	Greater than 1 km

Table 11.5. Transmission types, data rates, and maximum cable lengths.

Networks

A local-area network, or LAN, is a computer network that spans a relatively small area such as a single building or group of buildings. Most local-area networks connect workstations, personal computers, and peripherals together. Each node, or individual computer, in a local-area network works independently but also has the capability to access data and devices anywhere on the network. As a result, many users can share peripheral devices as well as data.

Local-area networks can transmit data rates faster than the data rates for telephone transmissions. However, local-area networks have limitations in terms of the distance and the number of connected computers. To offset these limitations, one local-area network can connect to other local-area networks over any distance through telephone lines and radio waves. A system of local-area networks connected in this way make up a wide-area network, or WAN.

The following characteristics differentiate one network from another:

• *topology:* The geometric arrangement of devices on the network. For example, devices can be arranged in a ring or in a straight line.

• *protocols:* The rules and encoding specifications for sending data. The protocols also determine whether the network uses a peer-to-peer or client/server architecture.

• *media:* Devices can be connected by twisted-pair wire, coaxial cables, or fiber optic cables. Some networks do without connecting media altogether, communicating instead via radio waves.

ISDN and Digital Subscriber Line Technologies

An integrated services digital network, or ISDN, functions as an international communications standard for sending voice, video, and data over digital telephone lines or normal telephone wires. The ISDN BRI standard supports data transfer rates of 64 kbits per second. Because most telephone companies use two ISDN B channels for the simultaneous transmission of voice and data, the data transfer rate doubles to 128 kilobits per second. Later versions of the ISDN standard support transfer rates of 1.5 Mbits per second.

Digital Subscriber Line, or DSL, technologies use sophisticated modulation schemes to pack data onto copper wires. However, the DSL technologies function only between the telephone switching station and the home or office. Both DSL and ISDN technologies operate over existing telephone lines and require short cable runs of less than 20,000 to a central telephone office. Despite the limitations, DSL offers great potential as a distribution technology because of the capability to transfer downstream data at rates up to 32 Mbits/second and upstream data at 1 Mbit/second.

Table 11.6 compares the data transfer rates of the different services and standards.

Network Type	Media	Data Rate	Use
Analog telephone systems Ethernet	Twisted Pair Twisted Pair Coaxial Cable Fiber Optics	56 kbps 10 Mbps	Home and Business access Business and School
Fast Ethernet	Twisted Pair Fiber Optics	100 Mbps	Business and School
Gigabit Ethernet Cable Modem	Fiber Optics Coaxial Cable (downstream only)	1 Gbps 512 Kbps to 52 Mbps	Large Enterprise Home, Business, and School Access
ISDN BRI Access	Twisted Pair	64 Kbps	Home and Business
ISDN PRI Enterprise	T-1	1.544 Mbps	Medium and Large
DSL	Twisted Pair	512 Kbps to 8 Mbps	Home, Business, and Small Enterprise Access
Satellite	Broadcast (Downstream Only)	400 Kbps	Home and Small Enterprise Access

Table 11.6. Communications options.

We can place the capabilities of these systems against the bandwidth requirements of NTSC and HDTV signals for comparison. After compression, a single NTSC format video signal with stereo audio requires a transfer rate of approximately 4 to 6 Mbits/second to ensure the reception of a high quality synchronized signal. Video in the HDTV format requires a transfer rate of approximately 15 to 20 megabits per second to provide a high-resolution reproduction of a televised scene. The transmission of 200 channels using the HDTV format would require a transfer rate of 4 Gbits/second.

Satellite Distribution of Data Services

Satellite reception dramatically increases the available bandwidth for communications to the consumer and provides an instantly accessible communications infrastructure. As an example, a satellite with sixteen transponders has a downlink transfer capacity of 600 Mbits/second. However, the asymmetric characteristics of satellite communications limit the transfer capacity to the downlink communications because of the lack of bi-directional bandwidth capabilities. Upstream communications back to the satellite continue to require the use of a telephone system.

Receiving Data Services with DIRECPC

With DirectPC, a service offered by Hughes Network Systems that provides Internet access through private satellite dishes, upstream requests for World WideWeb pages travel through a normal modem connection to the DirecPC Network Operating Center. The installation of a DirecPC system requires the purchase of a receiving system, an adapter card for a personal computer, and a connection to an Internet Service Provider, or ISP. DirecPC provides the optimum results when installed on a Pentium-class computer system with 32 MB of RAM and the Windows 95 or higher operating system. *Figure 11.9* shows a diagram of the DirecPC system.

DirecPC offers several options for consumers that include the Internet-only DirecPC installation kit and the DirecDuo Internet and television reception kit. While the Internet-only kits include a single-function DirecPC dish antenna, a DirecPC satellite modem, installation and access software, and a universal mount, the DirecDuo kit includes a slightly different dish antenna. The satellite modem connects either through an Industry Standard Architecture (ISA) connector, the Peripheral Connect Interface (PCI) connector, or the Universal Serial Bus (USB) connector found either within the computer or as a port.

DirecPC software installs with Windows 3.1, Windows 95, Windows NT, or Windows 98 and provides access to the Turbo Internet, Turbo Webcast, Turbo Newscast services. The services provide a sports ticker that provides real-time delivery of scores of sporting events and a stock ticker that shows the status of all three stock exchanges. The Turbo Newscast service provides the real-time video stream delivery of news.

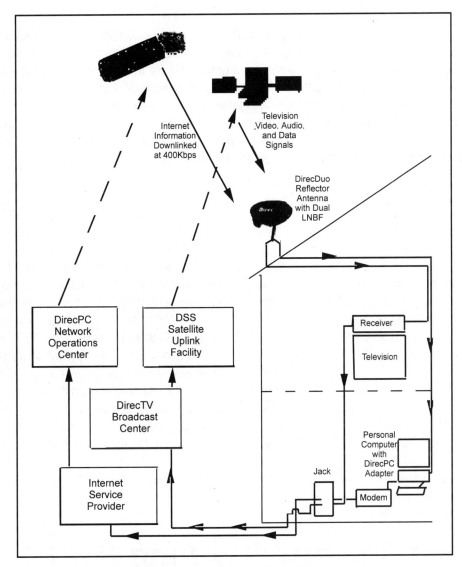

Figure 11.9. Diagram of the DirecPC system.

DirecPC Satellites

DirecPC relies on the GE-1 and GE-4 satellites placed in a geosynchronous orbit to carry and downlink the Internet data. Because GE-4 is an older satellite, systems aimed at GE-4 use the ISA rather than the PCI or USB satellite modems. In addition, the GE-1 Satellite relies on vertical polarization rather than the horizontal polarization seen with GE-4.

Along with the installation and applications software, the DirecPC it also arrives with a satellite alignment software program that lists the elevation and azimuth for most cities in the United States. After the consumer selects the appropriate city, the program responds with the exact elevation, azimuth, magnetic azimuth and polarization settings required for that location. In addition, the software includes a signal strength meter that shows as a graphical output to the screen and has a maximum strength reading of 160. The DirecPC dish antenna system orients to the correct satellite in the same way seen with the DBS systems in chapter eight.

DirecPC Operation

When a customer requests an address for a web page, the request travels from the customer's modem to the Internet Service Provider. As the customer sends the request, the DirecPC software attaches an electronic addressing mask called a tunneling code to the request. The tunneling code instructs the ISP to forward the request for the web page to the Network Operations Center instead of the ISP server. After the receiving the request, circuitry at the Network Operations Center strips the tunneling code from the request. Multiple T-3 lines carry the web page request to the appropriate web site where the Network Operations Center retrieves the desired content. The center uplinks the information to the DirecPC satellite. From there, the satellite downlinks the information at rate of 400 kilobytes per second to the customer's system and computer. With the entire transfer from computer to the Network Operations Center, the satellite, and back to the DirecPC system and personal computer taking approximately 450 milliseconds, the downloading of a 1 MB file takes approximately thirty seconds.

Installing and Running DIRECPC with Windows 95 or 98

Because the DirecPC software originally installed with Windows 3.1, the installation process involved manually installing networking components, software drivers, and a terminate-stay resident program. Furthermore, the Windows 3.1 operating system had base memory limitations that complicated the DirecPC installation. The installation of the networking components required that the installer have knowledge about hardware requirements and IRQ conflicts that can occur in a personal computer.

An interrupt request, or IRQ, line allows devices attached to the personal computer to send interrupt signals to the microprocessor. The interrupt switches the microprocessor from one operation to another temporary operation. Adding a new device to a personal computer requires the setting of the device IRQ number. IRQ conflicts occur when two devices have the same IRQ number. The plug-and-play function of Windows 95 allows the system to automatically set IRQ numbers and accept the installation of new devices such as interface cards.

With the introduction of Windows 95 and Windows NT, DirecPC offers a new set of installation software that uses the Win95 Winsock and allows the installation of other network cards. In brief, Windows 95 supports 32-bit applications and removes the memory and networking limitations formerly seen with DOS-based operating systems. The Win95 Winsock functions as an Application Programming Interface, or API, that allows communication between Windows programs and other machines through the TCP/IP protocol. The Transmission Control Protocol/Internet Protocol, or TCP/IP, represents a suite of communications protocols that connect host computers on the Internet. Network operating systems include protocols that support the TCP/IP standard.

Installing the DirecPC requires following a simple step-by-step process in Windows.

After placing the DirecPC disk in the floppy disk drive of the computer:

- Select the Start button at the bottom left of the Windows desktop,

- Select the Settings option from the list,

- Select the Control Panel icon,

- Select the Add/Remove Programs icon,

- Highlight DirecPC in the selection window,

- Select the Install button,

- At the "Install by Floppy or CD screen?" select Next,

- At the "Run Installation Program" screen, select Finish,

- At the Setup prompt, select Yes,

- Then, highlight "US" for Antenna location and select Next,

- At the Installation Path screen, select Next,

- At the "Select folder" menu screen, highlight the "DPC" selection and select Next,

- At the Choose Destination Location screen, select Next,

- Select Finish to complete the installation and restart the computer.

Gateways and Domain Name Servers

Use of the DirecPC service requires that the personal computer communicate through either a Serial Line Internet Protocol (SLIP) or the newer Point-to-Point (PPP) protocol so that a connection can occur between the computer, an ISP, and the DirecPC Network Operations Center. In addition, the installer must designate a gateway address in the network set-up portion of the Windows software. A gateway functions as a combination of hardware and software and links two different types of networks. The gateway number shown in the DirecPC installation designates a specific DirecPC satellite.

Setting up a network connection in Windows also requires the designation of a Domain Name Service, or DNS, number. A DNS translates the domain names associated with web addresses into IP addresses. Each individual IP address contains a unique set of numbers that point toward a specific Internet server. The installation of the DirecPC software specifies the DNS number (198.77.116.8) as a default setting. However, because of demands on the DirecPC server, using the DirecPC DNS number may slow the response of the system. Rather than use the DirecPC DNS, set the Windows designation for the DNS of the desired Internet Service Provider and set the DirecPC Navigator to use the modem for DNS lookups. The procedure includes the following steps:

- Locate the DNS designation for the ISP,

- In Windows 95, select the Start function. Then, select the Settings function and the Control Panel. Within the Control Panel, select the Network application.

- Select TCP/IP from the listed choices. After selecting TCP/IP, select DirecPC Adapter. Open the properties selection for the DirecPC adapter and select the DNS tab.

- Delete any previously set DNS numbers.

- Set the ISP DNS number as the default and select OK.

- Before restarting the computer, initialize the DirecPC Navigator application and select the Networking option.

- Select the Terrestrial Tab and follow the instructions for setting up a port to access the terrestrial only page using the DNS setting.

Restart the computer so that all changes record in the system.

To run DirecPC, select the Internet icon. Then, select properties and de-select auto dial. Starting the DirecPC TurboInternet application automatically dials the desired ISP. After logging onto the system, use either an Internet browser to access the World Wide Web.

Using DIRECPC on a Network

A computer system using DirecPC can function as a proxy server on a local-area network. Proxy server software acts as a gateway to the Internet for all client computers connected to the network as well as enabling the DirecPC capabilities on the network. Because a proxy server requires a high bandwidth connection, the server software resides on the computer using the DirecPC adapter card.

Before installing the proxy server software, attach the DirecPC computer to a functioning TCP/IP-based network. Install the DirecPC computer to the network and verify the compatibility of the DirecPC interface and the network interface card. Both cards must function at the same time on the network. The DirecPC computer will have TCP/IP settings configured for the DirecPC interface card. When configuring the TCP/IP settings on the DirecPC computer for network access, only change the TCP/IP settings for the network interface card.

Summary

Chapter eleven served as a capstone for the Howard W. Sams Guide to Satellite Technology. The chapter showed several methods for expanding and enhancing satellite reception systems. In addition, the chapter discussed different methods for achieving mobility with a satellite system. The chapter concluded by showing how to use a satellite system to gain access to the Internet.

As with other chapters in this book, chapter eleven built on information covered in previous chapters. The opening section about adding receivers to a system used procedures first seen in chapters five, six, and seven. Within that discussion, the chapter also provided more detailed descriptions of items such as line amplifiers, modulators, and compensators. The section that covered the installation of a DBS system in a motor home relied on techniques seen in chapters seven and eight. When the chapter moved to the discussion about satellite systems and Internet access, it pulled from knowledge gained throughout the text.

As a whole, the Howard W. Sams Guide to Satellite Technology provides a thorough overview of the theory that makes satellite communications possible and the practical issues surrounding the implementation of the technology. The coverage of the electronic systems that make up the receivers, positioners, and decoders combines with information about C-band, Ku-band, and DBS systems. All this provides the information needed to make a wise purchase and to install a system.

Index

Symbols

feedback 36, 41, 65, 141, 144, 164
feedback circuit 38
feedhorn 55, 86, 93, 110, 111
feedhorn/amplifier assembly 80
fiber optics 283, 284, 285
fiberglass 77, 78
fiberglass dishes 77
fiberglass reflectors 94
field-effect transistors 39, 143
figure of merit 73
filter 10, 139
filter capacitors 49, 134, 141, 150, 151, 155
filter choke 134
filter resistor 135
filtering 141, 142, 155
fishing vessels 62
Fixed Satellite Service band 192
fixed transponder bandwidth 225
flag information 229
flameproof resistors 141
flip-flops 23
fluorescent lights 72
FM 42
FM demodulator 174
focal length 65, 67
focal point 66, 67, 68
footprint 83
forward error correction 195, 232
frame rate 218
frame scanning rate 243
frame-by-frame analysis 229
free space loss 84
frequency 24, 41, 46, 72, 164, 249
frequency bands 33
frequency bandwidth 241
frequency divider 265
frequency division 161
frequency domain 26
frequency interleaving 244
frequency modulated waveform 45
frequency modulation
    42, 44, 45, 48, 49, 173, 177
frequency multiplication 161
frequency multiplier 161, 265
frequency response curve 61
frequency selectivity 235
frequency spectrums 24, 234

frequency synthesis 161, 162, 163, 164, 166, 168
frequency synthesizer control units 168
frequency synthesizers 162
FSS band 192
full-wave bridge rectifier circuit 130
full-wave bridge rectifiers 132, 142
full-wave rectification 132
full-wave rectifier 131, 132
full-wave rectifier circuit 130, 139

G

gain 36, 39, 40, 48, 68, 74, 76, 239
gas 113
gas storage tanks 62
gates 22, 33
General Instrument Corporation 208
General Instruments 261
geostationary orbital radius 67
geostationary satellites 103
geosynchronous arc 93
geosynchronous orbit satellites 64
geosynchronous orbits 52, 193, 289
geosynchronous satellites 35, 52, 83, 97, 104, 193
germanium 12
Globalstar 53
gravel 111
gray PVC 113
grid resistor 246
ground pole 110, 112, 113
ground pole installation 111
ground wave 27
grounding the satellite system 115
grounding the transmission cable 116
guide data 195
guy wires 110

H

half-wave circuits 131
half-wave rectifier 130, 131, 133
half-wave rectifier circuit 130, 134
Hall effect sensors 120, 121
harmonics 33, 243
HBO 51
HDTV 208, 209, 211, 213, 214, 215, 227, 287

white raster 245
wide-area network 286
wide-area networks 283
wind loading 107, 112, 124
wind loading conditions 118
Windows 3.1 288, 290
Windows 3.1 operating system 290
Windows 95 288, 291
Windows 98 288
Windows NT 288, 291

Z

zener diode 12, 137, 152
zener diode regulation 155
zener diode regulators 137
Zenith Electronics 208

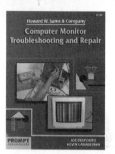

PUBLICATIONS

Disk Included!

Disk Included!

RadioScience Observing
Volume 1
by Joseph Carr

Among the hottest topics right now are those related to radio: radio astronomy, amateur radio, propagation studies, whistler and spheric hunting, searching for solar flares using Very Low Frequency (VLF) radio and related subjects. Author Joseph Carr lists all of these under the term "radioscience observing" — a term he has coined to cover the entire field.

In this book you will find chapters on all of these topics and more. The main focus of the book is for the amateur scientist who has a special interest in radio. It is also designed to appeal to amateur radio enthusiasts, shortwave listeners, scanner band receiver owners and other radio hobbyists.

RadioScience Observing also comes with a CD-ROM containing numerous examples of radio frequencies so you can learn to identify them. It also contains detailed information about the sun, planets and other planetary bodies.

Communications Technology
336 pages • paperback • 7-3/8 x 9-1/4"
ISBN: 0-7906-1127-9 • Sams: 61127
$34.95

RadioScience Observing
Volume 2
by Joseph Carr

Joe Carr expands on Volume 1 with all-new material, covering: techniques and methods; hardware design and construction; more RadioScience theory; and related geo-science and planetary science activities.

The discussions are more skill-oriented, and the design and construction more specific, building on the material presented in Volume 1. Again, a CD-ROM is included, containing information for making calculations, as well as audio clips, enabling readers to actually experience the thrills of "listening to the heavens."

Chapter topics include: Propagation Anomalies; Electromagnetic Fields and Safety; Using Used TVRO Dishes in RadioScience Observing; Finding Compass Bearings; Seismic Observations; Ultraviolet and Ozone Measurements; Receiver Selection; HF Low-Noise Amplifiers; Radio Telemetry on a Budget; Setting Up for SETI; and much more!

Communications Technology
440 pages • paperback • 7-3/8 x 9-1/4"
ISBN: 0-7906-1172-4 • Sams: 61172
$34.95

**To order your copy today or locate your nearest Prompt®
Publications distributor : 1-800-428-7267 or www.hwsams.com**
Prices subject to change.

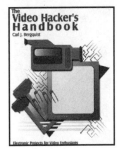

PROMPT®
PUBLICATIONS

IC Design Projects
Stephen Kamichik

Build Your Own Test Equipment
Carl J. Bergquist

IC Design Projects discusses some of the most popular and practical ICs, and offers you some projects in which you can learn to create useful and interesting devices with these ICs.

Once you have read through this book and completed some of its projects, you will have a real, working knowledge of ICs, enabling you to design and build you own projects!

Topics include: how power supplies operate, integrated circuit voltage regulators, TTL logic, CMOS logic, how operational amplifiers work, how phase-locked loops work, and more!

Projects include: battery charger, bipolar power supply, capacitance meter, stereo preamplifier, function generator, DC motor control, automatic light timer, darkroom timer, LM567 tone decoder IC, electronic organ, and more!

Test equipment is among the most important tools that can be used by electronics hobbyists and professionals. Building your own test equipment can carry you a long way toward understanding electronics in general, as well as allowing you to customize the equipment to your actual needs.

Build Your Own Test Equipment contains information on how to build several pragmatic testing devices. Each and every device is designed to be highly practical and space conscious, using commonly-available components.

Projects include: Prototype Lab, Multi-Output Power Supply, Signal Generator and Tester, Logic Probe, Transistor Tester, IC Tester, Portable Digital Capacitance Meter, Four-Digit Counter, Digital Multimeter, Digital Function Generator, Eight-Digit Frequency Counter, Solid-State Oscilloscope, and more.

Projects
261 pages • paperback • 7-3/8 x 9-1/4"
ISBN: 0-7906-1135-X • Sams 61135
$24.95

Professional Reference
267 pages • paperback • 7-3/8 x 9-1/4"
ISBN: 0-7906-1130-9 • Sams: 61130
$29.95

To order your copy today or locate your nearest Prompt®
Publications distributor : 1-800-428-7267 or www.hwsams.com

Prices subject to change.

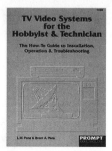